ENVIRONMENTAL GEOLOGY

FACING THE CHALLENGES OF OUR CHANGING EARTH

JON ERICKSON

FOREWORD BY PETER D. MOORE, PH.D.

Facts On File, Inc.

ENVIRONMENTAL GEOLOGY
Facing the Challenges of Our Changing Earth

For information contact:

Facts On File, Inc.
132 West 31st Street
New York NY 10001

Library of Congress Cataloging-in-Publication Data

Erickson, Jon, 1948–
Environmental geology : facing the challenges of our changing earth / Jon Erickson ; foreword by Peter D. Moore.
p. cm.—(The living earth)
Includes bibliographical references and index.
ISBN 0-8160-4727-8 (acid-free paper)
1. Environmental geology. I. Title.

QE38.E75 2002
550—dc21 2001051108

Facts On File books are available at special discounts when purchased in bulk quantities for businesses, associations, institutions, or sales promotions. Please call our Special Sales Department in New York at 212/967-8800 or 800/322-8755.

You can find Facts On File on the World Wide Web at http://www.factsonfile.com

Text design by Cathy Rincon
Cover design by Nora Wertz
Illustrations by Jeremy Eagle

Printed in the United States of America

VB Hermitage 10 9 8 7 6 5 4 3

This book is printed on acid-free paper.

CONTENTS

TABLES

ACKNOWLEDGMENTS

The author thanks the National Aeronautics and Space Administration (NASA), the National Oceanographic and Atmospheric Administration (NOAA), the National Park Service, the U.S. Army, the U.S. Army Corps of Engineers, the U.S. Department of Agriculture (USDA), the USDA Forest Service, the USDA Soil Conservation Service, the U.S. Department of Energy, the U.S. Geological Survey (USGS), and the U.S. Navy for providing photographs for this book.

The author would also like to thank Frank K. Darmstadt, Senior Editor, and Cynthia Yazbek, Associate Editor, for their invaluable contributions to the development of this book.

FOREWORD

Now more than 6 billion people live in the world, and each one of us need air to breathe, food to eat, water to drink, and space to live in. So how many more of us can the Earth support? Double the present population? Well, maybe.

The trouble is that the Earth is constantly changing, so trying to make forward projections regarding the resources available and our future demand for them becomes difficult. Some of the changes we see taking place around us are entirely natural, and perhaps unavoidable, but many others result from the consequences of human activities. If we are to understand the hopes and the hazards facing humankind, then we need to know more about the causes—and the consequences—of environmental change. Here the science of environmental geology comes to the rescue.

The Earth has always been restless. The shifting crust creates volcanoes, and fault lines shuffle erratically, giving rise to earthquakes. The uneven distribution of the Sun's energy over the face of the Earth creates the turbulence of winds that can reach destructive force. We might regard these as unfortunate hazards, but a planet without such energetic motion would be a dead planet. We need to learn to live alongside these great forces, perhaps even to harness them and use them for our own benefit, and only environmental geology can teach us how.

To learn that our activities on the planet can actually have an impact on the great forces of nature has come as a great surprise to us. Only 50 years ago scientists would smile complacently at the idea that by burning fossil fuels, such as coal and oil, we could affect the proportion of carbon dioxide in the atmosphere. Surely, the great reservoirs of the globe, the atmosphere, the oceans, and the soils, could absorb any wastes that our single species could pour into them? Now we know that we are capable of upsetting the balance of nature and that we are in the process of doing so. The great cycles of the earth, carbon, nitrogen, water, that are described here, need to be understood in much greater quantitative detail if our impact is to be properly appreciated and subsequently reduced.

We now know that the Earth's temperature is steadily rising, and it is very likely that human activity is contributing to this process. The consequences of this rise, for wildlife and for us, are still being debated. Climate will undoubtedly change, perhaps leading to greater instability in the weather of certain parts of the world. Some areas already under intense human pressure, such as the arid regions, may become even less productive than they are already, leading inevitably to human hunger and suffering. Some species of plants and animals might be pushed to the brink of extinction, and we shall lose their unexplored genetic resources. Fisheries, forestry, and agriculture will all be affected by the coming changes, resulting in considerable social repercussions.

The study of environmental geology might not yet be able to provide all of the answers to these problems, but at least it supplies us with a greater clarity of vision. The aim of this book is to provide that vision. The need for information and understanding becomes ever greater as we see the changes taking place around us. Change has always been with us, but at no time in history has the rate of change been so rapid. Facing the problems that confront us on our changing Earth requires informed minds and a new way of thinking. This book seeks to give us the information we need, but it will also stimulate imaginative thoughts and new ideas, which are needed if the coming generation is to meet the challenges and maintain the quality of our planet.

—Peter D. Moore, Ph.D.

INTRODUCTION

Environmental geology is a relatively new science. It deals with the relationship between people and their geologic environment. The main emphasis is on natural hazards and the ecological problems created by human activities. Environmental geology applies geologic information toward solving natural and human-caused problems. Other topics of importance to environmental geology include ecology, the interaction of life and its habitat; hydrology, the cycle of water from the sea onto the land; natural resources, the sources of minerals and energy; and land use, the utilization of Earth's land surface.

The Earth is always changing. These changes result from natural processes, but many are also caused by human activities. High population growth with its rising demands on the environment and increasing pollution is in the process of transforming the planet in a manner comparable to the effects of long-term geologic processes. Consequently, people have been called the *human volcano* because human influences on the environment are global, similar to major volcanic eruptions. Humans therefore constitute a major geologic force on the face of Earth.

The text begins by examining the forces of nature, how they affect living conditions on Earth, and the cycles that maintain life. It then investigates air and water pollution, waste disposal, and the restoration of the environment. Next, it discusses how greenhouse gases and air pollution can change the climate. It then shows how the climate works toward distributing water over the land, the effects and control of flooding to save people and their property. This is followed by a discussion about what happens to river-borne sediments once they reach the ocean, the effects coastal processes have on people living near shore, and how the sea reclaims the land.

The text continues with an examination of the effects of ground shaking and volcanic activity and their dangers to society. Next, it concentrates on the destructive forces of erosion, slides, and ground failures. It then takes a look at the geologic hazards of desert regions and the role that droughts, advancing deserts, desertification, and roving sand dunes play in people's lives. It then deals with the depletion of natural resources and explores other forms of energy. Finally, it examines the importance of proper management of the land for the preservation of all life.

Science enthusiasts will particularly enjoy this fascinating subject and gain a better understanding of how the forces of nature operate on Earth. Students of geology and earth science will also find this a valuable reference to further their studies. Readers will enjoy this clear and easily readable text that is well illustrated with dramatic photographs, detailed illustrations, and helpful tables. A comprehensive glossary is provided to define difficult terms, and a bibliography lists references by chapter for further reading. The geologic processes that shape the surface of our planet are examples of the tireless forces that make this a living Earth.

1

THE BALANCE OF NATURE

THE NATURAL PROCESSES

This chapter examines the forces of nature and how they affect living conditions on Earth. Many aspects of life are governed by cycles. Perhaps the periodicity of the Earth-Moon system was responsible for the initiation of life in the first place. The constant waxing and waning of the tides account for the prodigious growth in the intertidal zones. The Earth has its own inner cycles that affect geologic and biologic processes. The continuous evaporation of seawater and the precipitation on the continents is one of nature's most important cycles.

The rock cycle is responsible for volcanic activity, which has a profound influence on climate and life. The circulation of carbon required for maintaining the balance between incoming and outgoing thermal energy determines the temperature of the planet. Indeed, without the greenhouse effect, which traps heat in the atmosphere that otherwise would escape into space, nothing could live on this planet. The recycling of nitrogen in the biosphere is also fundamental for the support of living beings on Earth.

THE BIOSPHERE

The biosphere, which comprises all living entities, is more extensive and contains more life than previously thought possible. Life on the Earth's surface is so apparent we often forget that most of the world's living beings lie hidden well out of sight. These are simple organisms involved in nutrient recycling, which helps sustain all other forms of life.

Since life's first humble beginnings, it has responded to a variety of chemical, climatological, and geographical changes in the Earth, forcing species either to adapt or to become extinct. Many dead ends along branches of the evolutionary tree are found in the fossil record, which itself represents only a fraction of the species that have actually lived. Nearly every conceivable form and function of organisms have been tried, some more successful than others. Through this trial and error method of specialization, natural selection has chosen some species to prosper while others perish.

Species have adapted to nearly every conceivable environment—from subfreezing to boiling, strongly acidic to toxic alkaline—and to extreme pressures of the abyss and far below the ocean floor. Living bacteria have also been discovered deep underground. Bacterial spores trapped in amber millions of years ago and buried under thick sediments have miraculously been brought back to life. On the bottom of the ocean in the cold and dark lies an eerie world occupied by some of the strangest creatures on the planet. The discovery of complex animals, sometimes called extremeophiles, living within such unexpected and bizarre habitats shows how resilient life is even under extreme conditions.

Very few places on Earth are truly devoid of life. It is found in the hottest deserts and the coldest polar regions. It resides in the lowest canyons and tallest mountains. Life also exists in the deepest oceans and the highest regions of the troposphere. Nor is life excluded from scalding-hot springs (Fig. 1) or high-temperature environments deep below the ground. Although species most frequently encountered on Earth's surface seem to be the most dominant force in shaping the planet, the unseen microbes actually do the most work. They comprise about 90 percent of the biomass, the total weight of all living matter. These are morphologically simple creatures that are biochemically diverse, highly adaptive, and absolutely necessary for maintaining living conditions on Earth.

Single-celled photosynthetic organisms thriving in the sunlit zone of the ocean generate about 80 percent of the atmospheric oxygen. Microorganisms such as bacteria play a critical role in breaking down the remains of plants and animals to recycle nutrients in the biosphere. Surface plants depend on bacteria in their root systems for nitrogen fixation. Bacteria live symbiotically in the gut of animals and aid in the digestion of food. Biologic processes are responsible for massive concentrations of silicon, carbon, iron, manganese, copper,

Figure 1 *The largest hot springs in the western group of Snake River Hot Springs, Yellowstone National Park, Wyoming.*

(Photo by J. D. Love, courtesy USGS)

and sulfur in Earth's crust. Simple organisms also comprise the bottom of the food chain on which life ultimately depends for its survival.

Life constitutes a major geologic force that has made extensive changes to Earth. The evidence of biospheric processes in Earth's history belongs to the broad field of biogeology. Stromatolite structures (Fig. 2) appear to be the earliest fossilized remains of microorganisms, which existed as far back as 3.5 billion years ago. The 3.8-billion-year-old carbonaceous sediments of the Isua Formation in southwest Greenland show a depletion of carbon 13 with respect to carbon 12, which is thought to be a common manifestation of biologic activity. Therefore, life processes appear to have been operating for at least 80 percent of Earth's history.

With this much time involved, life has brought about some dramatic and extensive changes to the Earth. The first major alteration came with the deposition of banded iron formations on continental margins by iron-metabolizing bacteria. These formations are mined extensively for iron ore around the world. The second was the conversion of most of the carbon dioxide in the atmosphere and ocean into oxygen when photosynthesis evolved as an energy source (Fig. 3). The oxygen generated by photosynthesis produced a secondary benefit, namely the ozone layer in the upper stratosphere. The shielding of the Sun's harmful ultraviolet rays by the ozone layer made conditions safe for land plants and animals to cover the Earth, which constituted a major change.

Another dramatic change came when humans evolved. The burning of tremendous amounts of fossil fuels, the pollution of the environment with

Figure 2 *Stromatolite structure near the junction of Canyon Creek and Salt River, Gila County, Arizona.*

(Photo by A. F. Shride, courtesy USGS)

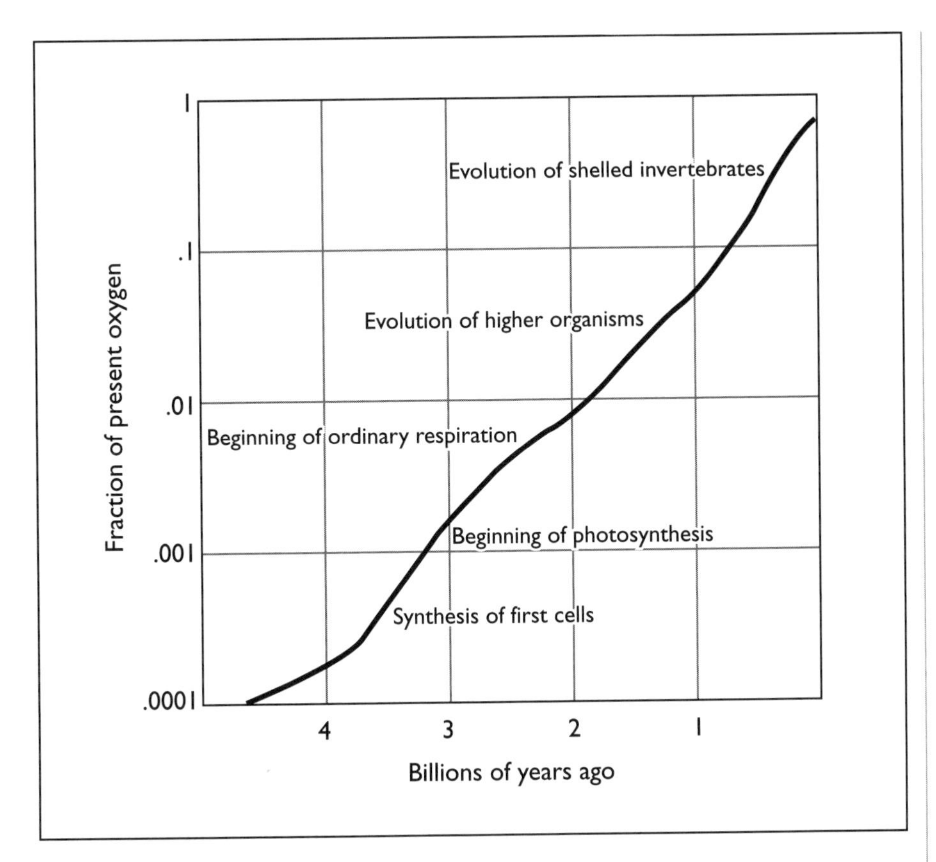

Figure 3 *The evolution of life and oxygen.*

toxic wastes, the destruction of forests and wildlife habitats, and the uncontrollable human population explosion place humans in the unique position of causing major changes to Earth in a comparatively short period. This makes humans a major biogeologic force on the face of the planet.

THE GAIA HYPOTHESIS

The Earth not only gives life all the necessities for survival, but life also appears to have made some of its own changes to maintain itself at optimal levels. The Gaia hypothesis, named for the Greek goddess of the Earth, suggests that life, to some extent, controls its environment to optimize living conditions. It portrays the planet as a single huge living organism that creates a favorable environment for itself. The interaction between the living and nonliving thus establishes a self-regulating system that maintains a constant state of equilibrium.

Life defined in physical terms is a huge, intricate molecular machine that seems to overcome, at least temporarily, the second law of thermodynamics, which essentially states that every form of order eventually dissolves into chaos. Life manages to go against the flow of a steadily decaying universe. However, the uphill struggle comes at the expense of a great deal of energy supplied by the Sun. This energy is manifested by the presence of large amounts of oxygen in the atmosphere and ocean. Without life, chemical reactions would have slowed down, and all oxygen would have long ago been bound to other elements in the crust.

As life progressed, slow but steady changes took place that greatly affected the outcome of the planet. Like Earth, the other planets and their satellites have a core, a mantle, a crust, and even an atmosphere or a liquid or icy hydrosphere. However, only Earth has a biosphere. Moreover, the biosphere requires more than just having living entities. Life must also be integrated with the lithosphere, hydrosphere, and atmosphere to constitute a fully developed biosphere.

Largely because of plate tectonics, life was able to flourish on this planet. Possibly, active plate tectonics could not operate if Earth did not possess life as well. Lime-secreting organisms in the ocean remove carbon dioxide, an important greenhouse gas, from the atmosphere and store it in the bottom sediments. This keeps Earth's surface temperature within the range needed for plate tectonics to operate effectively, which in turn maintains living conditions on the planet.

Originally, the atmosphere contained about 25 percent carbon dioxide, or roughly the same percentage as the amount of oxygen that exists today. Because the Sun's output was about a third less than at present, the high levels of carbon dioxide helped to maintain Earth's temperature. Without this stabilizing gas, the planet would have completely frozen over, and because ice is such a good reflector of sunlight, the Earth would have remained an icy orb. Indeed, during the worst period of glaciation around 680 million years ago, even the Tropics froze.

When green plants evolved, they gradually replaced carbon dioxide with oxygen via photosynthesis, which manufactures organic compounds and produces oxygen as a by-product. Meanwhile, the Sun was becoming progressively hotter. Large amounts of atmospheric carbon dioxide were no longer needed. If carbon dioxide had not been removed from the atmosphere by the biosphere, Earth could have suffered the same fate as Venus, whose high surface temperatures evaporated its oceans eons ago. Moreover, if Earth had begun with the atmosphere it has today, it would have been as cold as Mars. Either way, life could not have survived.

At first, simple organisms lived in an anaerobic (lacking oxygen) environment, in which oxygen was poisonous to life. When photosynthesis first evolved, as early as 3.5 billion years ago, when algae built the first stromatolites, all oxygen that was being produced by plants bonded to chemical elements. This oxygen was permanently locked up in Earth's crust. About 2 billion years ago, these oxygen sinks held all the oxygen they could contain. The gas began to build up slowly in the ocean and atmosphere. As the oxygen content reached higher levels, complex organisms began to evolve (Table 1). When the level approached present-day amounts, the ozone screen enabled plants and animals to conquer the land.

Life appears to maintain oxygen and carbon dioxide in a perfect balance. Too much of one with respect to the other could have disastrous consequences. Life-forms use the atmosphere both as a source of raw materials, such as oxygen and nitrogen, and as a repository for waste products, such as carbon dioxide. In this manner, life is directly linked to the greenhouse effect. Living organisms can therefore regulate the climate to their own benefit. Thus, without life, Earth's climate would be wildly out of control.

THE PRECARIOUS BALANCE

A fortunate set of circumstances has held global temperatures to within the freezing and boiling points of water. Earth's distance from the Sun has a major effect on the temperature range, as a distance difference of only 10 percent could spell the difference between life or death on this planet. Even minor changes in orbital variations can initiate the onset of ice ages.

The climate is most significantly influenced by the greenhouse effect (Fig. 4), which traps solar energy that would otherwise escape into space. If Mars had Venus's heavy carbon dioxide atmosphere, it would be hotter than Earth despite being farther out in the solar system. A strong greenhouse effect

TABLE 1 EVOLUTION OF LIFE AND THE ATMOSPHERE

Evolution	Origin (million years)	Atmosphere
Origin of Earth	4,600	Hydrogen, helium
Origin of life	3,800	Nitrogen, methane, carbon dioxide
Photosynthesis	2,300	Nitrogen, carbon dioxide, oxygen
Eukaryotic cells	1,400	Nitrogen, carbon dioxide, oxygen
Sexual reproduction	1,100	Nitrogen, oxygen, carbon dioxide
Metazoans	700	Nitrogen, oxygen
Land plants	400	Nitrogen, oxygen
Land animals	350	Nitrogen, oxygen
Mammals	200	Nitrogen, oxygen
Humans	2	Nitrogen, oxygen

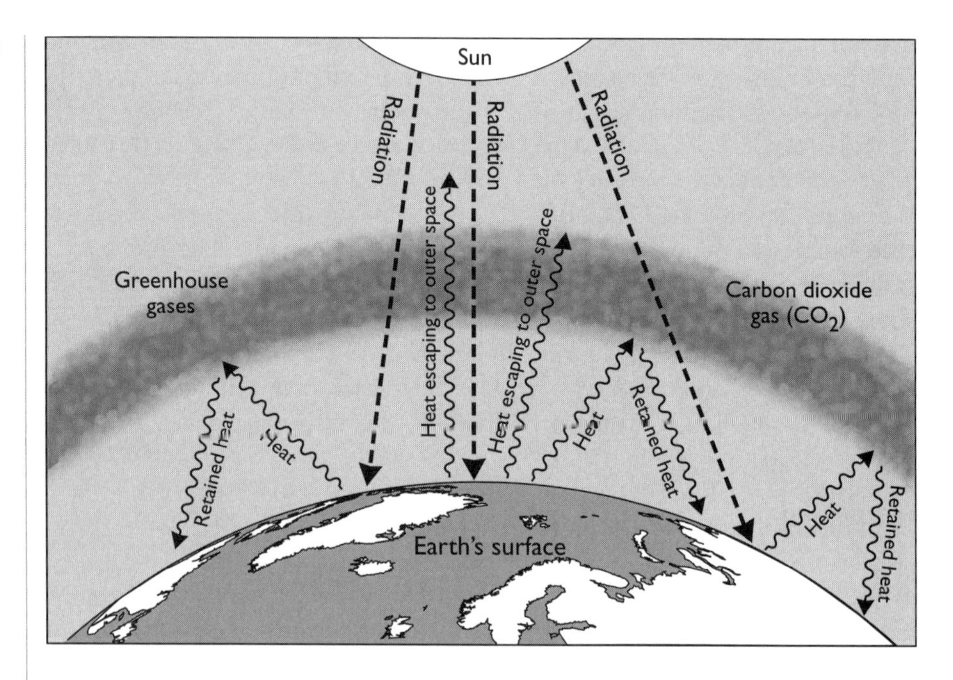

Figure 4 *The principle of the greenhouse effect. Incoming solar energy striking the surface is converted to infrared radiation (IR) that is trapped by atmospheric greenhouse gases and reradiated back to the ground.*

would retain what little heat Mars receives from the Sun—heat normally lost to space. On the other hand, if Venus, the planet closer to the Sun, had Mars's thin carbon dioxide atmosphere, it would be colder than Earth.

Life therefore owes its existence to the greenhouse effect. Large quantities of greenhouse gases in the early atmosphere maintained temperatures within tolerable, life-sustaining limits, even though the Sun's output was lower than it is today. Fluctuations in the carbon dioxide content of the atmosphere have influenced major changes in the world's climate. During the ice ages, when the carbon cycle removed large quantities of carbon dioxide from the atmosphere, temperatures plummeted and great ice sheets flowed across the land. When vigorous volcanic activity added excessive amounts of carbon dioxide to the atmosphere, temperatures soared. Earth became a hothouse. Only when carbon dioxide levels remain uniform does the climate maximize benefits for all life-forms.

The ecosphere, which integrates life with other Earth processes, provides living beings with all the essentials needed for survival. Life also might have made major changes of its own to maintain optimum living conditions as suggested in the Gaia hypothesis. The biosphere, the portion of Earth in which life exists, appears to be able to control the environment to some extent by regulating the climate. This is similar to how the human body regulates its temperature to optimize metabolic efficiency. For example, a certain species of plankton releases into the atmosphere a sulfur compound that aids in cloud formation. If

Earth warms, plankton growth is invigorated. This releases more cloud-forming sulfur compounds to cool the planet and thereby stabilize temperatures.

Photosynthetic organisms store energy by combining carbon from atmospheric carbon dioxide with hydrogen from water to form carbohydrates. Coal deposits are essentially buried solar energy because they originated as lush vegetative matter in ancient swamps, as evidenced by fossilized stems and leaves (Fig. 5). Vast subterranean reservoirs of petroleum are basically cooked hydrocarbon molecules from once-living microorganisms. These fossil fuels have been accumulating over millions of years. When they are burned in factories, furnaces, and vehicles, the equation reverses. Carbon is recombined with oxygen, releasing carbon dioxide back into the atmosphere (Fig. 6).

Humans are therefore interfering with the carbon cycle by spewing massive quantities of carbon dioxide into the atmosphere from industrialization and habitat destruction including deforestation and loss of wetlands. The combustion of tremendous amounts of fossil fuels, the pollution of the environment with toxic wastes, the destruction of the forests, the extinction of species, and the uncontrolled human population explosion places humans in a unique position of inflicting major changes to Earth in a comparatively short period. In this manner, people are fast becoming the single most destructive force on the face of the planet, confounding nature's efforts to maintain the balance.

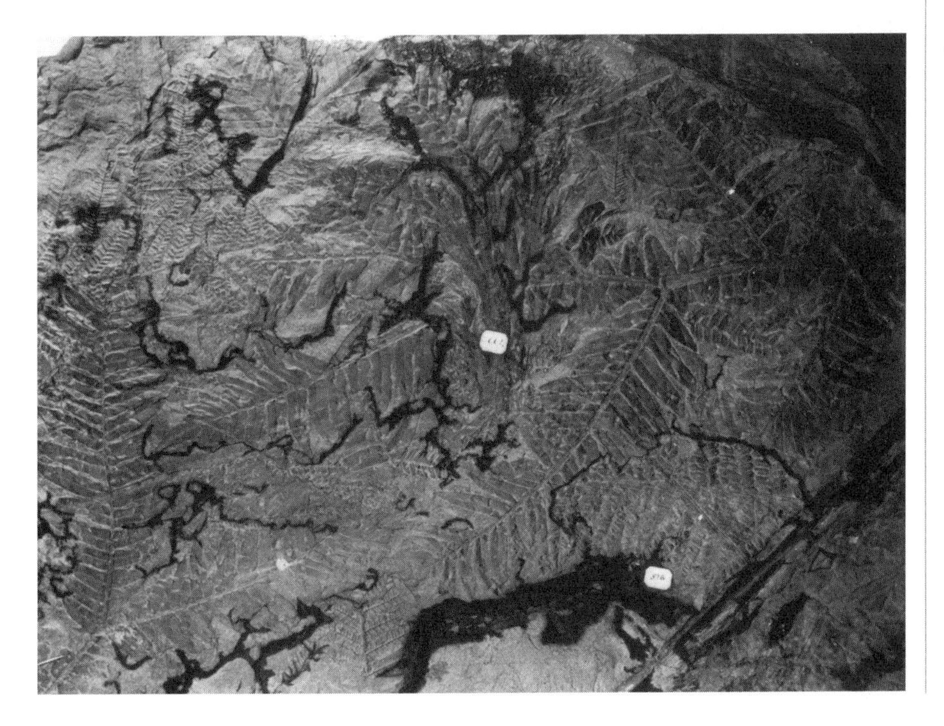

Figure 5 *Fossil leaves of the tree fern* Neuropteris, *Fayette County, Pennsylvania.*

(Photo by E. B. Hardin, courtesy USGS)

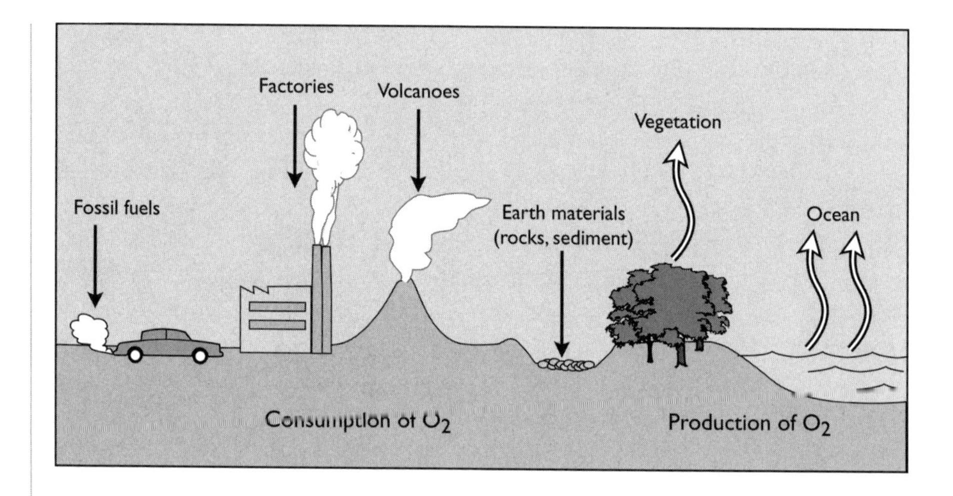

Figure 6 *Oxygen consumed by the combustion of fossil fuels and the oxidation of earth materials is balanced by plant respiration on land and sea.*

THE ENERGY BUDGET

The atmosphere maintains a balance between incoming solar radiation and outgoing infrared radiation. Earth intercepts about 1 billionth of the Sun's rays. Only about half this solar energy reaches the surface, where 90 percent evaporates water. This releases heat energy to the atmosphere when water vapor condenses into clouds. Earth reradiates out into space the same amount of energy it receives from the Sun; otherwise, temperatures would become excessively hot. However, if Earth emitted too much heat, temperatures would turn extremely cold. This delicate balancing act is known as the energy or heat budget (Fig. 7). It is responsible for maintaining global temperatures within the narrow confines where life is possible.

When sunlight strikes Earth's surface, it transforms into infrared energy. This is absorbed by the atmosphere and emitted to space from altitudes between 15,000 and 20,000 feet. The solar energy striking the ground averaged over a year and spread evenly around the world is more than 1 million watts for an area equal to the size of a football field. This amount is about 5,000 times greater than the energy radiating from Earth's interior.

The angle that sunlight strikes the surface also determines the amount of solar energy being absorbed or reflected. In the Tropics along the equator, the Sun's rays strike Earth from directly overhead. Therefore, more solar radiation is absorbed on the surface than is reflected into space. In the polar regions, the Sun's rays strike Earth at a low angle. Thus, more solar radiation is reflected into space than is absorbed on the surface. If not for the distribution of heat by the atmosphere and ocean, the Tropics would swelter in heat and the higher latitudes would shiver in cold. This would occur to such a degree that few places on Earth would be inhabitable.

Solar energy also scatters sideways due to dust particles and aerosols in the atmosphere from such sources as dust storms, forest fires, sea salt, meteors, air pollution, and volcanoes—the greatest source of natural air pollutants (Fig. 8). These fine atmospheric particles are responsible in large part for making the sky blue. This color lies at the high range of the solar spectrum and is scattered by the atmosphere. If the atmosphere did not disperse light, the daytime sky would be as black as night, and the Sun would appear as a very large star. When the Sun is low on the horizon, its rays must pass through so much atmosphere that only the red colors can penetrate, producing crimson sunrises and sunsets.

The heat budget is also responsible for generating the weather. Warm air rises at the equator in narrow columns and travels aloft toward the poles. In the polar regions, the air liberates heat, cools, sinks, and returns to the equator, where it warms again in a continuous cycle. This energy exchange is actually conducted by three convective loops called Hadley cells, named for the British meteorologist George Hadley, the discoverer of atmospheric convection. Currents in the ocean act in a similar manner, only more slowly, taking much longer to complete the journey. The middle latitudes, or temperate zones, become battlegrounds between warm, moist tropical air and cold, dry polar air. When these air masses clash, they create storms.

The distribution of air masses is also responsible for the world's winds. The Coriolis effect bends the air currents in response to Earth's rotation. A point on the surface moves faster at the equator than near the poles because

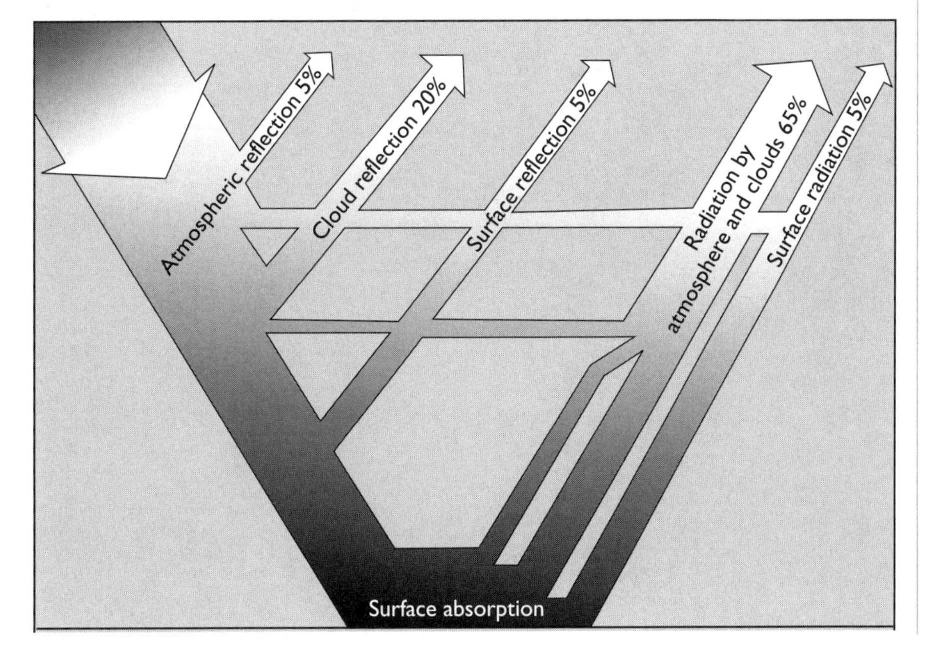

Figure 7 *Earth's heat budget.*

it is farther from the axis of rotation and therefore must travel a greater distance in a given time. Air currents moving toward the poles deflect to the east because the ground beneath them is slowing. Air currents moving toward the equator deflect to the west because the ground beneath them is accelerating.

The oceans play a vital role in distributing solar energy. Radiation from the Sun heats seawater. Solar energy is transported by ocean currents; lost by conduction, radiation, and evaporation; and regained by precipitation (Fig. 9). Heat flow between the oceans and atmosphere is responsible for cloud formation. A tremendous amount of thermal energy is required to evaporate seawater

Figure 8 *Eruption of Cerro Negro Volcano, west-central Nicaragua, in November 1968.*

(Photo courtesy USGS)

Figure 9 The heat balance of Earth involves the evaporation of seawater into clouds, which radiate energy during precipitation, and the distribution of heat by ocean currents.

into water vapor. When the clouds move to other parts of the world, they liberate energy by precipitation, which helps circulate heat around the globe.

Another transfer mechanism from the sea to the air is the transport of marine-borne substances by the wind. These materials are ejected into the atmosphere by bursting air bubbles and ocean spray from waves. The fine spray evaporates into small particles of sea salt wafted aloft by air currents. Upward of 10 billion tons of salt enter the atmosphere in this manner annually. The salt also provides seed crystals for the condensation of rain.

The oceans are also responsible for the steady onshore and offshore breezes. During the day, the land warms to a higher degree than the sea. Warm air rises from the land and travels aloft toward the sea, where it cools and descends landward. At night, the land cools below the temperature of the sea. Warm air rises from the sea and travels aloft toward the land, where it cools and descends seaward. The monsoon winds, which provide life-sustaining rains to many regions of the world, operate in much the same way, except they are seasonal phenomena.

The heat budget mostly depends on the albedo effect (Fig. 10 and Table 2). This is an object's ability to reflect sunlight and is dependent on color and texture. Some things reflect solar energy better than others because of their greater reflective properties. Light-colored objects, such as clouds, snowfields, or deserts, reflect more solar energy than they absorb. Dark-colored objects, such as oceans or forests, absorb more solar energy than they reflect. Most solar

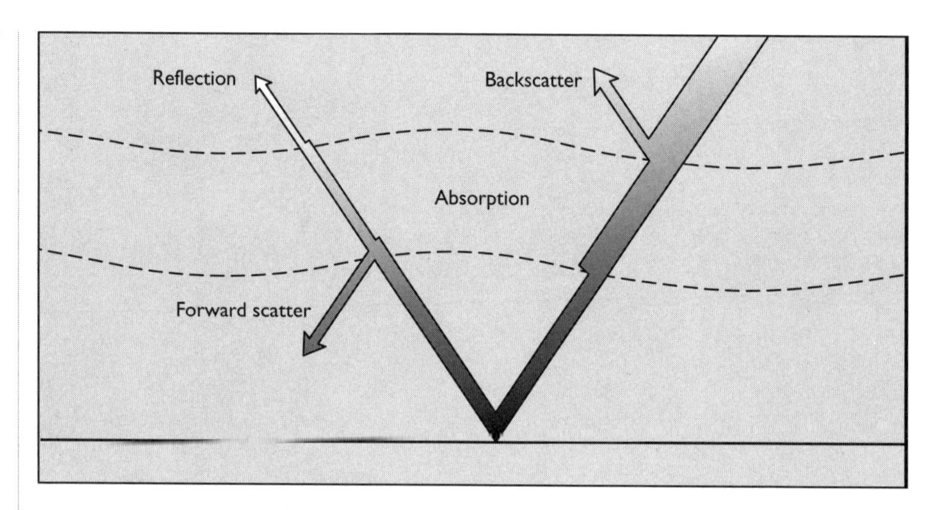

Figure 10 *The effect of the albedo on incoming solar radiation.*

energy impinging on the ocean evaporates seawater. This energy is lost to space when water vapor condenses into rain.

Fully one-third of the solar energy is reflected back into space before it has a chance to heat Earth. Most of this lost energy is reflected off clouds. Data

TABLE 2 ALBEDO OF VARIOUS SURFACES

Surface	Percent reflected
Clouds, stratus	
< 500 feet thick	25–63
500–1,000 feet thick	45–75
1,000–2,000 feet thick	59–84
Average all types and thicknesses	50–55
Snow, freshly fallen	80–90
Snow, old	45–70
White sand	30–60
Light soil (or desert)	25–30
Concrete	17–27
Plowed field, moist	14–17
Crops, green	5–25
Meadows, green	5–10
Forests, green	5–10
Dark soil	5–15
Road, blacktop	5–10
Water, depending upon sun angle	5–60

from satellites indicate that, on the whole, clouds exert a net cooling influence on the planet. The effect is much stronger at midlatitudes than in the Tropics. High cirrus clouds (Fig. 11) retain Earth's heat. In contrast, low stratus clouds block out the Sun and cool the surface.

Figure 11 *Different cloud types reflect or absorb the Sun's energy.*

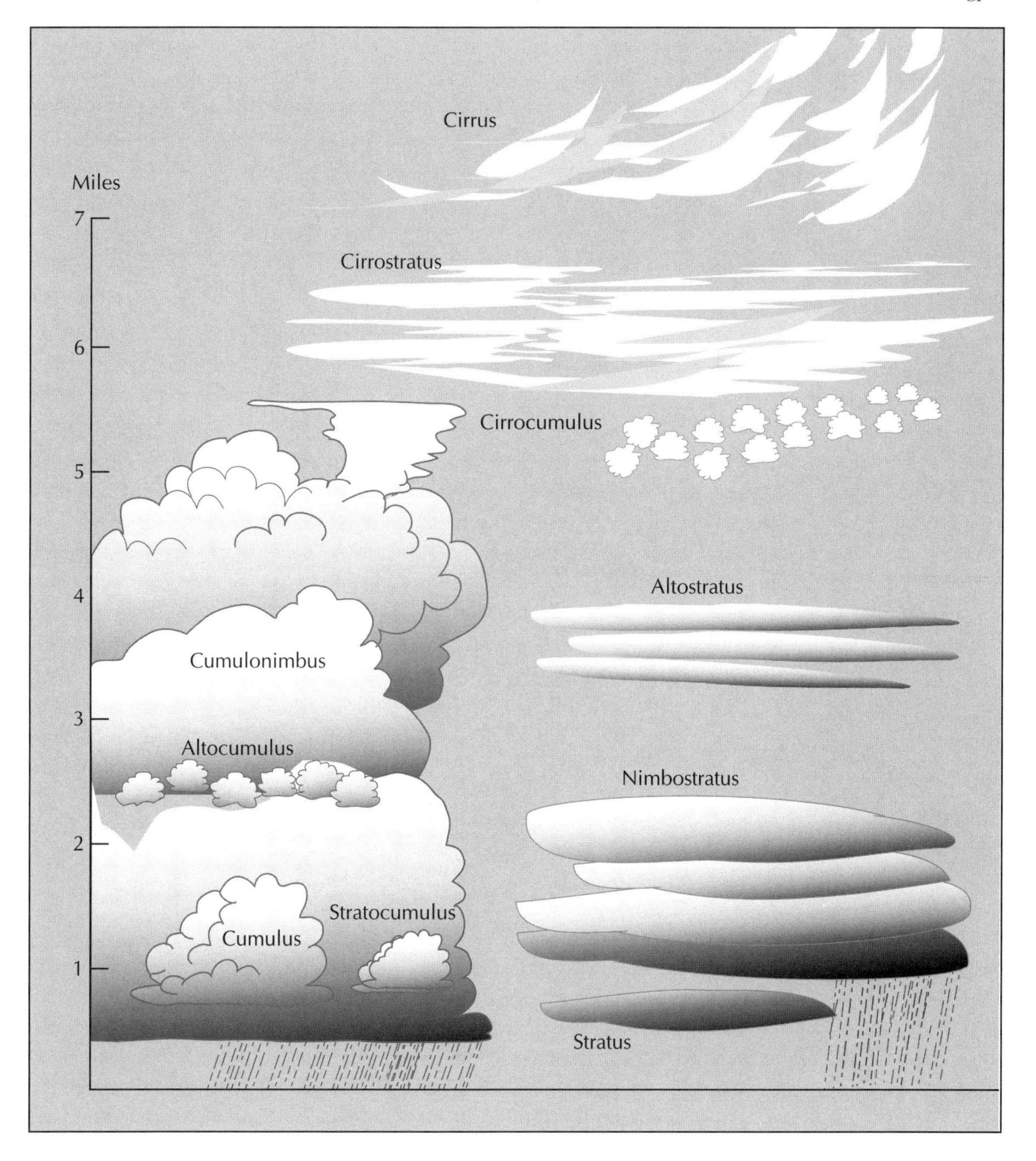

As Earth warms, the increased cooling ability of clouds moderates the heating. Over the Tropics, the heating and cooling effects of clouds nearly balance out. Over the midlatitudes, clouds do the most cooling. Fine solid or liquid particles injected into the atmosphere by natural sources or by man-made pollutants also block the Sun's rays from reaching the ground. These particles do allow infrared heat rising from the surface to escape into space, causing a net cooling of Earth.

Only about half the total solar energy reaches the surface to heat the ocean and the land. On land, most of this energy is absorbed by soil and plants. Vegetation use the red and blue hues for photosynthesis but have no need for green light, so it is reflected away. This is what gives plants their green color. Eventually, all sunlight striking the surface converts into infrared energy and radiates upward. If not for the greenhouse effect to prevent all the infrared from departing Earth, this would indeed be a very cold planet.

THE OCEAN CURRENTS

The ocean's ability to store and transport vast quantities of heat is much more effective than the atmosphere's and has a profound effect on the climate. The ocean's large heat capacity allows it to retain summer's heat and release it during winter, thereby moderating Earth's temperature during the seasons. Every summer, the ocean surface warms by as much as 15 degrees Celsius above its preceding winter value. About one decade is required to change the temperature of the upper 1,000 feet of the ocean significantly and thousands of years for the temperature of the entire ocean to change. This is called the oceanic thermal lag. The ocean's heat capacity is so large that up to 100 years or more are required to respond fully to global climate change.

Surface and abyssal currents in the ocean move heat around the planet (Fig. 12). The ocean surface currents are driven by steady winds and function similar to currents in the atmosphere. They transport warm water from the tropics, distribute it to the higher latitudes, and return with colder water. As with flowing air masses, surface currents in the ocean are deflected by the Coriolis effect to the right in the Northern Hemisphere and to the left in the Southern Hemisphere.

Abyssal currents are driven by thermal forces in the ocean. Cold water in the polar regions descends, spreads out upon hitting the ocean floor, heads toward the equator, and rises in the tropics. The upwelling of deep seawater in the tropics also transports a high concentration of dissolved carbon dioxide, which produces a sharp carbon dioxide peak over the equator. The path taken by the deep-water currents is influenced by the distribution of landmasses and by the topography of the ocean floor. Abyssal currents flowing toward the

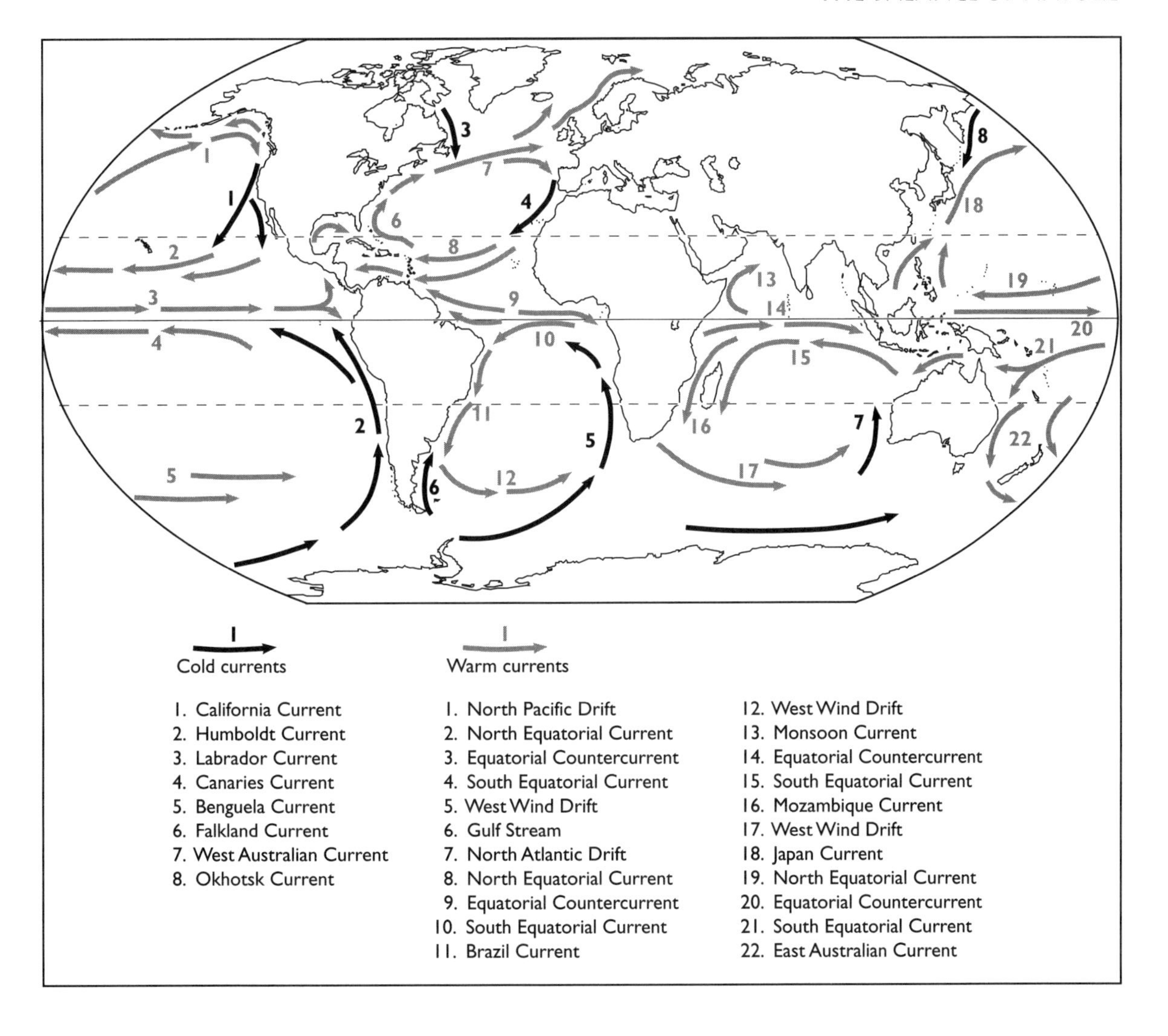

Figure 12 *The major ocean currents distribute heat around Earth.*

equator are deflected to the west due to Earth's eastward rotation, which presses the currents against the eastern edges of the continents.

The cold, dense, salty surface waters of the Arctic sink to the bottom and form a deep-sea current called the North Atlantic Deep Water (NADW). This is a subsurface ocean river whose volume is 20 times greater than the combined flow of all the world's rivers on land. Another subsurface current called the western boundary undercurrent, which flows along eastern North America, transports some 20,000 cubic miles of water yearly.

Water sinking in the polar regions is matched by upwelling currents in the Tropics. This creates an efficient heat transport system that completes the journey from the tropics to the poles and back again in upward of 1,000 years.

Upon reaching the Tropics, the cold water of the abyss rises toward the surface. These upwelling currents play an important role in transporting deep ocean-bottom nutrients to the surface to support marine life. Although these zones cover only about 1 percent of the ocean's surface area, they sustain about 40 percent of all marine life.

When parts of the ocean currents become pinched off, they form eddies or rings of swirling water that play an important role in mixing the ocean waters. They are up to 100 miles or more across and reach as deep as 3 miles below the surface. Marine life often gets trapped in these rings and is transported to hostile environments, where they survive only as long as the rings continue to operate, usually several months.

Ocean currents have a dramatic effect on the weather. Changes in these systems can send abnormal weather patterns around the world. Once about every two to seven years, anomalous atmospheric pressure changes in the South Pacific, called an El Niño Southern Oscillation (Fig. 13) causes the westward-flowing trade winds to collapse. Warm water piled up in the western Pacific by the winds then flows back to the east, creating a great sloshing of water in the South Pacific Basin. The layer of warm water in the eastern Pacific becomes thicker. This suppresses the thermocline, the boundary between cold- and warm-water layers, and prevents the upwelling of cold water from below. This temporarily disrupts the upwelling of nutrients, which adversely affects the local marine biology.

The opposite condition results during a La Niña, when the surface waters of the Pacific cool. In mid-1988, water temperatures in the central

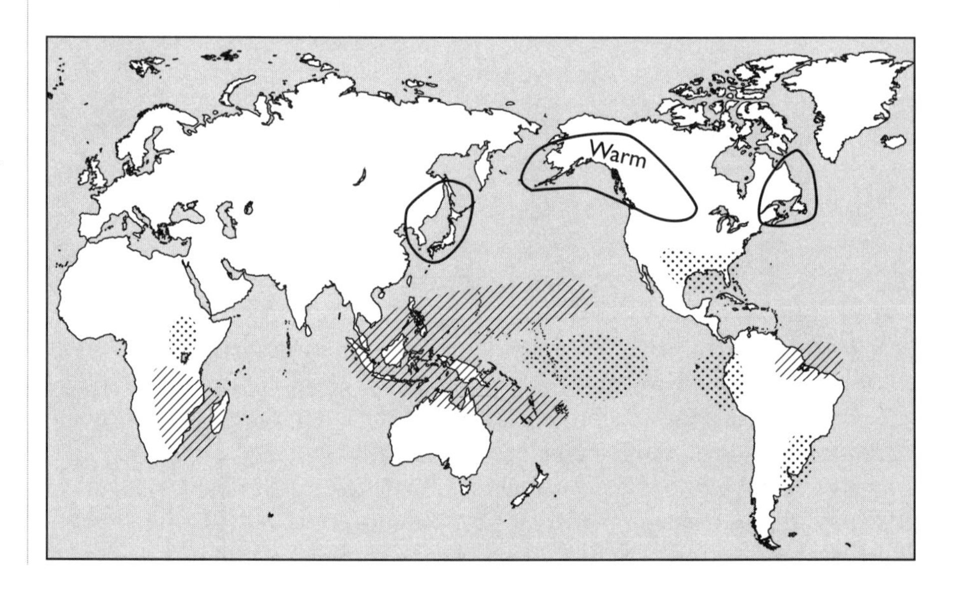

Figure 13 *Typical northern winter temperature and precipitation patterns during El Niño warming in the Central Pacific. Hatched areas are dry, stippled areas are wet, and encircled areas are warm.*

Pacific plummeted to abnormally cold levels, signaling a climate swing from an El Niño to a La Niña. Strong monsoons hit India and Bangladesh, and heavy rains visited Australia during that year. The La Niña might also have been responsible for a severe drought in the United States during 1988 and a marked drop in global temperatures in 1989. The unusually strong El Niño of 1993 was responsible in large part for the Midwest floods, the costliest in the nation's history. The lack of either an El Niño or a La Niña in the winter of 2000–01 might have been responsible for record cold conditions in many parts of the country.

Ocean waves are produced by the wind blowing across the surface of the water. Storms at sea create most waves that strike against nearby shores. Offshore hurricanes produce the largest waves, which can be extremely damaging. The wind-stirred layer of the ocean involves the upper 100 to 200 feet. It is the most uniform environment on Earth and is always in equilibrium with the atmosphere. However, it is only a thin film on the surface of an over 2-mile-deep ocean, the vast bulk of which is near freezing.

Ocean tides result primarily from the gravitational pull of the Moon on the Earth (Fig. 14). The Sun also raises tides in the ocean but to a lesser extent. The Moon revolves around Earth in an elliptical orbit and exerts a stronger pull on the near side of the planet than on the far side. The difference between the gravitational attraction on both sides is about 13 percent, which elongates the center of gravity of the Earth-Moon system. As Earth spins on its axis, the oceans flow into two tidal bulges. One faces toward the Moon, and the other faces away from it. The ocean is therefore shallower between the tidal bulges, giving it a slightly oval shape. The maximum high tide in the middle of the ocean rises only about 2.5 feet. However, due to the motion of the sea and the geography of the coastline, tides often rise several times higher.

Earth's daily rotation causes each point on the surface to go into and out of two tidal bulges. Thus, the tides appear to rise and fall twice daily. The Moon also orbits Earth in the same direction it rotates, creeping ahead a little every day. By the time a point on Earth's surface has rotated halfway around, the tidal bulges have moved forward with the Moon, and the point must travel farther each day to catch up. Therefore, the actual period between high tides is 12 hours 25 minutes.

The maximum tidal amplitude occurs twice monthly during the new and full moon, when the Earth, Moon, and Sun align in nearly a straight-line symmetry known as syzygy, from the Greek *syzygos* meaning "yoked together." This configuration causes spring tides, from the Saxon word *sprignam* meaning a rising or swelling of water. The minimum tidal amplitude produces neap tides during the first and third quarters of the Moon, when the Earth, Moon, and Sun align at right angles to one another and when the solar and lunar tides oppose each other.

The waxing and waning of ocean tides are responsible for the prodigious growth in the intertidal zone (Fig. 15), the habitat between high and low tides. The pounding surf shapes the activity patterns of inhabitants living on beaches exposed to the open sea. Intertidal organisms occupying protected bays are not as exposed to the ocean's fluctuations. They are controlled instead by more subtle conditions such as a drop in temperature or pressure changes induced by the incoming tides.

Most marine life lives within the mixed layer of the ocean, the top 250 feet, called the phototropic zone. These marine organisms must live near the surface where sunlight can penetrate for photosynthesis. The surface action of the ocean plays an important role in the exchange of carbon dioxide and oxygen, of which 80 percent of Earth's total supply is generated by marine plants. However, if none of the oxygen were removed by respiration and

Figure 14 *Earthrise over the lunar horizon from an Apollo spacecraft.* (Photo courtesy USGS)

Figure 15 *Intertidal zone near Piller Point, Clallam County, Washington.*

(Photo by W. O. Addicott, courtesy USGS)

decay, its level could double in about 10,000 years, and Earth would incinerate itself.

THE GEOCHEMICAL CARBON CYCLE

The recycling of carbon through the geosphere makes Earth unique among planets. This is evidenced by the fact that the atmosphere contains large amounts of oxygen. Without the carbon cycle, this oxygen would have long since been buried in Earth's crust. Fortunately, plants replenish oxygen during photosynthesis, which plays a critical role in the circulation of carbon in the biosphere and therefore provides the basis for all life.

The geochemical carbon cycle (Fig. 16) is the transfer of carbon within the ecosphere. It involves interactions between the crust, ocean, atmosphere, and life. Carbon dioxide converted into bicarbonate is washed off the land and enters the ocean. Marine organisms convert it into carbonate sediments, which are thrust into Earth's interior and become part of the molten magma. Carbon dioxide then returns to the atmosphere by volcanic eruptions. Many aspects of this important cycle were understood around the turn of the 20th century, notably by American geologist Thomas Chamberlain and chemist Harold Urey. However, only in the last few years has the geochemical carbon cycle been placed within the more comprehensive framework of plate tectonics, responsible for geologic activity on Earth.

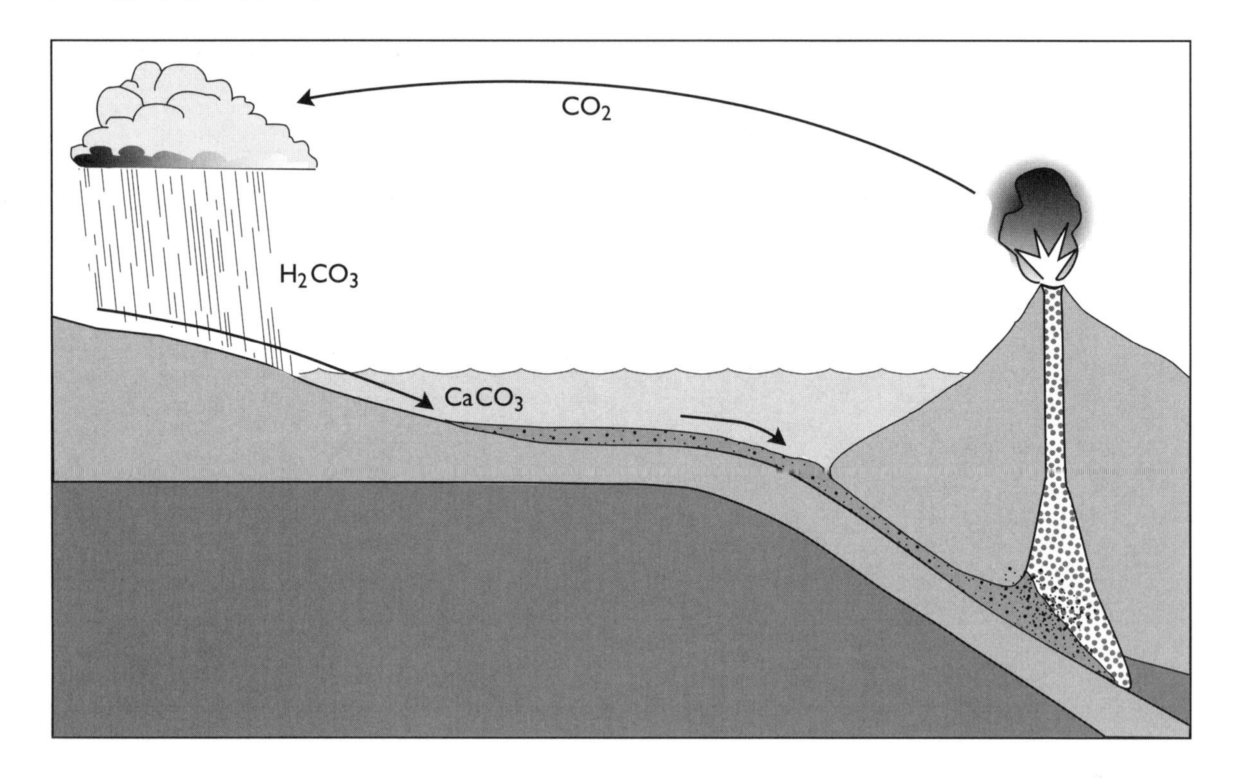

Figure 16 The geochemical carbon cycle. Carbon dioxide converted into bicarbonate is washed off the land and enters the ocean, where marine organisms convert it into carbonate sediments, which are thrust into the Earth's interior and become part of molten magma. Carbon dioxide is returned to the atmosphere by volcanic eruptions.

The biological carbon cycle is only a small component of this cycle. It is the transfer of carbon from the atmosphere to vegetation by photosynthesis to produce organic compounds. It returns carbon to the atmosphere when plants respire or decay. Only about one-third of the chemical elements, mostly hydrogen, oxygen, carbon, and nitrogen, comprising most of the elements of life, are recycled biologically.

The biosphere plays a very important role in the cycling of carbon. The creation and decomposition of peat bogs might have been responsible for most of the changes in levels of atmospheric carbon dioxide during the past two glaciations. The bogs have accumulated upward of 250 billion tons of carbon in the last 10,000 years since the end of the last ice age, mostly in the temperate zone of the Northern Hemisphere. Over geologic time, progressively more land has drifted into latitudes where large quantities of carbon are stored as peat. During the last million years, glaciations have gradually remolded large parts of the Northern Hemisphere into landforms more suitable for peat bog formation in wetlands.

The vast majority of carbon is not stored in living tissue but locked up in sedimentary rocks. Even the amount of carbon contained in fossil fuels is meager by comparison. Nevertheless, the combustion of large quantities of fossil fuels and the destruction of the world's forests is transferring more carbon to the atmosphere than it can dispose of. The burning of carbon-based fuels could dramatically influence the climate through the greenhouse effect. Because the man-made release of carbon dioxide is so much faster than natural processes, humans are short-circuiting the carbon cycle.

Carbon dioxide presently comprises about 365 parts per million of the air molecules in the atmosphere, amounting to about 800 billion tons of carbon. It is one of the most important greenhouse gases, which trap solar heat that would otherwise escape into space. Carbon dioxide, therefore, operates somewhat like a thermostat to regulate the temperature of the planet. Since it plays such a critical role in regulating Earth's temperature, major changes in the carbon cycle could have profound climatic effects. If the carbon cycle removes too much carbon dioxide, Earth cools. If the carbon cycle generates too much carbon dioxide, Earth warms. Therefore, even slight changes in the carbon cycle could considerably affect the climate.

The world's oceans play a vital role in regulating the level of atmospheric carbon dioxide. In the upper layers of the ocean, the concentration of gases is in constant equilibrium with the atmosphere. The mixed layer of the ocean contains as much carbon dioxide as the entire atmosphere. The gas dissolves into the waters of the ocean mainly by the agitation of surface waves. If the ocean did not contain photosynthetic organisms that absorb dissolved carbon dioxide, much of its reservoir of this gas would escape into the atmosphere, more than tripling the present content.

A large portion of the carbon in the ocean comes from the land. Atmospheric carbon dioxide combines with rainwater to form carbonic acid. The

Figure 17 *A quarry in Bangor limestone near Russellville, Alabama.*
(Photo courtesy USGS)

acid reacts with surface rocks to produce dissolved calcium and bicarbonate, which are carried by streams to the sea. Marine organisms use these substances to build their calcium carbonate skeletons and other supporting structures. When the organisms die, their skeletons sink to the ocean bottom, where they dissolve in the deep abyssal waters. The huge abyss contains the largest reservoir of carbon dioxide, holding 60 times more carbon than the entire atmosphere.

Sediments on the ocean floor and on the continents store most of the carbon. In shallow water, carbonate skeletons from once-living organisms build thick deposits of carbonate rock such as limestone (Fig. 17). The limestone permanently buries carbon in the crust. The burial of carbonate in this manner is responsible for about 80 percent of the carbon deposited onto the ocean floor. The carbon locked up in carbonate minerals in the upper crust is estimated at 800 trillion tons. The remaining carbonate originates from the burial of dead organic matter washed off the continents.

In this respect, marine life acts as a pump to remove carbon dioxide from the atmosphere and the ocean's surface waters and stores it in the deep sea. The faster this biologic pump works, the more carbon dioxide that is removed from the atmosphere, with the rate determined by the amount of nutrients in the ocean. A reduction of nutrients slows the biologic pump, returning deep-sea carbon dioxide to the atmosphere.

Half the carbonate transforms back into carbon dioxide, which returns to the atmosphere, mostly by upwelling currents in the tropics. Therefore, the concentration of atmospheric carbon dioxide is highest near the equator. If not for this process, in a mere 10,000 years, all carbon dioxide would be removed from the atmosphere. The loss of this important greenhouse gas would result in the cessation of photosynthesis and the extinction of life.

The final stage of the carbon cycle is the return of carbon to the environment by volcanic activity (Fig. 18). Volcanoes play an important role in restoring the carbon dioxide content of the atmosphere. The carbon dioxide escapes from carbonaceous sediments that melt in Earth's interior to provide new magma for volcanoes. The molten magma along with its content of carbon dioxide rises to the surface to feed magma chambers beneath volcanoes. When the volcanoes erupt, carbon dioxide is released from the magma and returns to the atmosphere.

THE NITROGEN CYCLE

The atmosphere is composed of 78 percent nitrogen, comprising a molecule of two atoms. Nitrogen is also a major constituent of living matter. Carbon,

Figure 18 *Molten lava pouring into the sea from a Hawaiian volcano.*

(Photo courtesy National Park Service)

nitrogen, and hydrogen are the essential elements for manufacturing proteins and other biological molecules. Nitrogen is practically an inert gas, however, and requires special chemical reactions before it can be used by nature. Therefore, to make nitrogen combine with other substances, a significant amount of energy is required.

Atmospheric nitrogen originated from early volcanic eruptions and the breakdown of ammonia, a large constituent of the primordial atmosphere comprising a molecule of one nitrogen atom and three hydrogen atoms. Unlike most other gases, which have been replaced or permanently stored in the crust, Earth retains much of its original nitrogen. This is because life prevents all nitrogen from transforming into nitrate, which is easily dissolved in the ocean, where denitrifying bacteria return the nitrate-nitrogen to its original gaseous state. Without this process, all the nitrogen in the atmosphere would have long ago disappeared. Earth therefore would be left with only a fraction of its present atmospheric pressure.

The nitrogen cycle is a continuous exchange of elements between the atmosphere and biosphere. It is spurred by the action of organisms such as nitrogen-fixing bacteria, which metabolize soil nitrogen for use by plants. These microbes remove nitrogen from the air and convert it into nitrogen

Figure 19 *Tube worms, large clams, and giant crabs are sustained by hydrothermal vents on the deep-ocean floor.*

compounds, which are incorporated into the tissues of plants or animals that feed on them. All methods of nitrogen fixation, which converts nitrogen into useful chemicals, require a source of abundant energy, mainly supplied by the Sun. Earth also provides a source of energy in the form of hydrothermal vents on the deep-ocean floor, which support the world's strangest biology (Fig. 19). The decay of organisms after death releases nitrogen back into the atmosphere, thus completing the cycle.

Human activities, however, have doubled the rate at which nitrogen gas in the atmosphere is chemically converted into compounds that can be used by plants and animals. The excess nitrogen has disrupted one of the planet's fundamental cycles, resulting in worldwide biological havoc. High nitrate levels can cause life-threatening diseases. Leaching of nitrates can seriously contaminate groundwater aquifers, streams, and even the seas. The increase in nitrogen compounds in the atmosphere results in the development of ozone destroyers, greenhouse gases, and pollutants contributing to urban smog. These compounds are also leaching soils of nutrients, increasing the acidification of surface waters, and clogging the seacoasts with nitrogen-hungry algae, which choke off other aquatic life.

The principal cause of this disruption is the rapidly growing global use of nitrogen fertilizers for agriculture, amounting to 40 percent of all nitrogen taken up by crops. In the middle 20th century, affluent nations accounted for more than 90 percent of all fertilizer consumption. However, by the end of the century, developing countries used more than 60 percent of the global output of nitrogen fertilizer. Often, nitrogen fertilizer is overused. Instead, fertilizer should be applied only where it is absolutely needed. The best means of reducing the growth in nitrogen use is finding more efficient ways

Figure 20 *Rafts of blue-green algae lying in a pool in Indian Canyon, Duchesne County, Utah.* (Photo by W. H. Bradley, courtesy USGS)

to fertilize crops. The nitrogen balance can also be returned to normal by restoring nitrogen-trapping wetlands.

The nitrogen oxides in acid rain can be especially harmful to aquatic organisms. Nitrogen works like a nutrient, promoting the growth of algae (Fig. 20).

Figure 21 *Barley that is about two weeks from harvesting, Howard County, Maryland.*

(Photo by T. McCabe, courtesy USDS Soil Conservation Service)

The algae block out sunlight and deplete water of its dissolved oxygen when the algae die and decay. This, in turn, suffocates other aquatic plants and animals. Widespread increases in nitrate levels along with higher concentrations of toxic metals, including arsenic, cadmium, and selenium, are occurring globally. The main factors contributing to this increase are fertilizer and pesticide runoff along with acid rain, which dissolves heavy metals in the soil. Acid rain also depletes the soil of its important nutrients, including calcium, magnesium, and potassium. Some soils are so acidic they can no longer be cultivated.

Crops require large amounts of fixed nitrogen. However, the natural supply of fixed nitrogen is finite, imposing a limit on world agriculture. Therefore, supplemental nitrogen must be supplied by chemical fertilizers if agriculture is to keep up with the ever-growing demand. Unfortunately, artificial fertilizers do not produce crops as nutritious as those grown using natural fertilizers. As a result, nations that regularly rely on chemical fertilizers could have inadequate diets.

Today's high-yield crops, including basic staples such as corn, wheat, and barley (Fig. 21), quickly deplete the soil of fixed nitrogen. This nitrogen must be replenished by applying either organic fertilizers—the preferred method—or chemical fertilizers, which require large amounts of energy to manufacture, usually supplied by fossil fuels. In this light, world agriculture along with the people it feeds could suffer catastrophically from a severe energy shortage.

After discussing the natural balance of Earth and the cycles that maintain life, the next chapter investigates how pollution affects the life processes.

2

ENVIRONMENTAL DEGRADATION

ECOLOGY AND POLLUTION

This chapter examines air and water pollution, waste disposal, and the restoration of the environment. Environmentalism advocates the preservation or improvement of the natural environment, especially conservation of Earth's resources, protection of wildlife habitats, and control of pollution. Environmental changes are occurring globally at rates never experienced before in this planet's recent history. Environmental degradation in large parts of the world appears to have surpassed the threshold of irreversibility, meaning its effects cannot be easily turned around. Pollution is a problem of worldwide consequences, requiring solutions on a global scale. It is so pervasive that more conservation can attack only part of the problem, not completely solve it.

The degradation of the environment, along with its accompanying threats to health and disruption of ecosystems, is nothing new, however. Human disturbance of the environment has been noted from earliest recorded history. What is different today is that pollution problems are becoming increasingly obvious, with subtle secondary reverberations that previously went unnoticed. Moreover, many environmental disturbances have begun to manifest themselves, altering the biosphere faster than at anytime in human history. Environmental catastrophes along with other disasters are becoming more commonplace as the world's population continues to expand.

ATMOSPHERIC POLLUTION

All foreign substances injected into the air from natural and artificial sources are defined as atmospheric pollution (Fig. 22). Atmospheric pollution has become a growing threat to health and welfare throughout the world because of the ever-increasing emissions of contaminants into the atmosphere (Table 3). The load of particulate matter consisting of soot and dust suspended in the atmosphere as the result of human activity is estimated at about 15 million tons and is rapidly escalating.

Natural pollutants include salt particles from ocean spray, pollen and spores released by plants, smoke from forest fires set by lightning strikes, wind-blown and meteoritic dust, and volcanic ash. Volcanoes are perhaps the largest natural polluters in the world. They produce sulfurous gasses that mix with

Figure 22 *Air pollution is a serious problem in some industrialized areas.*
(Photo courtesy NOAA)

TABLE 3 EMISSIONS OF CONTAMINANTS IN THE UNITED STATES (IN MILLIONS OF TONS PER YEAR)

Source	Carbon monoxide	Particulates	Sulfur oxides	Hydrocarbons	Nitrogen oxides
Transportation	92	1	1	12	10
Industry	9	12	29	13	15
Waste disposal	3	1		1	
Other	5	1		4	
Total	109	15	30	30	25

urban smog from nearby towns, greatly exacerbating local air pollution problems. Volcanoes also belch an impressive array of organic chemicals into the air. However, ozone-depleting halocarbons are insignificant compared with human-induced pollutants.

Air pollution is classified as either primary or secondary pollutants. Primary pollutants are emitted directly from principal sources such as factories and motor vehicles. Secondary pollutants originate from the chemical reactions among primary pollutants. Many reactions responsible for secondary pollutants are triggered by sunlight and are therefore called photochemical reactions. Nitrogen oxides produced by factory furnaces and motor vehicles absorb solar radiation and initiate a chain of complex chemical reactions. In the presence of organic compounds, these reactions form undesirable secondary products that are unstable, irritating, and toxic. Chief among these is surface ozone. This is not to be confused with stratospheric ozone, which forms a protective screen against solar ultraviolet radiation.

Ozone is a highly reactive gas that builds up in the lowest layer of the atmosphere, where it harms plants and animals. The pollutant forms when hydrocarbons and nitrogen oxides from the incomplete combustion of fossil fuels mix in the air and are "cooked" by sunlight. Surface ozone is responsible for about 90 percent of crop failures resulting from air pollution in the United States. Ozone pollution reduces crop yields by $5 billion to $10 billion annually. With China's rapidly expanding economy and population, its air pollution problems—particularly ozone—have reached the point where they are seriously harming agriculture. This raises questions about whether the world's fastest developing country can adequately feed itself in the future.

High ozone levels also reduce photosynthesis and increase cell damage in sequoia seedlings, dramatically reducing their survival rate. Giant sequoias were quite common in the past. Today, though, their range is confined to a 30-

mile-wide noncontinuous strip running some 500 miles along the Pacific coast from northern California to south of Big Sur. Loblolly pine growth rates in the American South have decline by as much as 15 percent because of air pollution, mostly surface ozone.

Most air pollution that reduces visibility, harms plants and animals, and corrodes man-made structures comprises dry deposits. These airborne particles consist of unburned carbon, dust particles, and minute sulfate particles. The finest particles, called aerosols, are produced mainly by chemical processes resulting from high-temperature combustion in coal-fired generating plants and internal combustion engines. These yield nitrogen oxide along with gaseous nitric acid.

The result of all these pollutants clogging the skies is a decrease in sunlight reaching the ground and a subsequent cooling of the surface. Sunlight striking airborne particulates also heats the atmosphere, causing a thermal imbalance and unstable weather. Possibly one reason an increased level of atmospheric carbon dioxide has not yet shown a substantial upward trend in global temperatures by the greenhouse effect is that much of it is offset by the cooling effects of particulate matter in the atmosphere.

Coarse atmospheric particles originate mainly from the mechanical breakup of naturally occurring substances such as ash from volcanic eruptions and sediment suspended in the air by dust storms. Large particles of carbon, or soot, are produced by agricultural fires, forest fires, and brushfires (Fig. 23) along

Figure 23 *A brushfire is used to burn the understory of a forest in Dallas County, Texas.*

(Photo courtesy USDA Forest Service)

with inefficient combustion of wood-burning stoves and fireplaces. In developing countries, cooking fires, vegetation burns, and industrial fuel combustion have spawned the buildup of airborne sooty particles or aerosols that result in a constant haze. Soot absorbs sunlight, heating the atmosphere and producing a temperature imbalance. This causes temperatures to rise with altitude—just the opposite of what they should do.

Large quantities of atmospheric soot generated by massive forest fires could result in abnormal weather throughout the world. Sooty, dark aerosols absorb sunlight and contribute to climate warming. Brighter aerosols such as sulfates and nitrates reflect sunlight back into space, exerting a cooling effect. In addition, the particles seed large, long-lasting clouds, which scatter light and promote cooling. The pollution also alters the cycle of rain and evaporation from the ocean, while precipitation returns to the surface as acid rain.

Slash-and-burn agriculture, which destroys millions of acres of forestland each year, is responsible for tremendous amounts of smoke entering the atmosphere. Dust blown off newly plowed or abandoned fields has been on the rise, clogging the atmosphere and severely eroding the land. Factory smokestacks and motor vehicle exhaust pipes send aloft huge quantities of soot and aerosols. Even in once pristine areas such as the Arctic tundra, significant levels of pollution create an Arctic haze originating from distant industrial sources to the south.

Geography also plays a major role in air pollution. Under an atmospheric inversion, with a warm layer of air acting as a lid on cold air near the ground, the smoke can create a persistent haze during winter. In areas such as the Los Angeles basin, polluted air settling in the valley does not escape during temperature inversions. The world's worst air pollution disaster occurred in 1952 during the great London smog, which killed some 4,000 people by a strong inversion lasting 11 days.

High-pollution days therefore do not necessarily indicate an increase in the output of pollution. Instead, the air into which the pollution is released is not disbursed by the wind, making the air more toxic. Stagnant air under a zone of high pressure allows little vertical mixing of the pollutants with the cleaner air aloft, and air quality drops precipitously. As global temperatures rise, a change in weather patterns, combined with sluggish movement of air masses, could produce long-lived, dirty air masses that hover over industrial centers.

One of the most disturbing surprises of the last few decades was the discovery of a huge hole in the ozone layer over Antarctica, where half the stratospheric ozone disappears during the southern winter (Fig. 24). When the ozone hole breaks up in the spring, ozone-depleted air travels to the midlatitudes, where ultraviolet levels climb appreciably. A similar ozone hole often hovers over the Arctic, jeopardizing parts of the Northern Hemisphere with

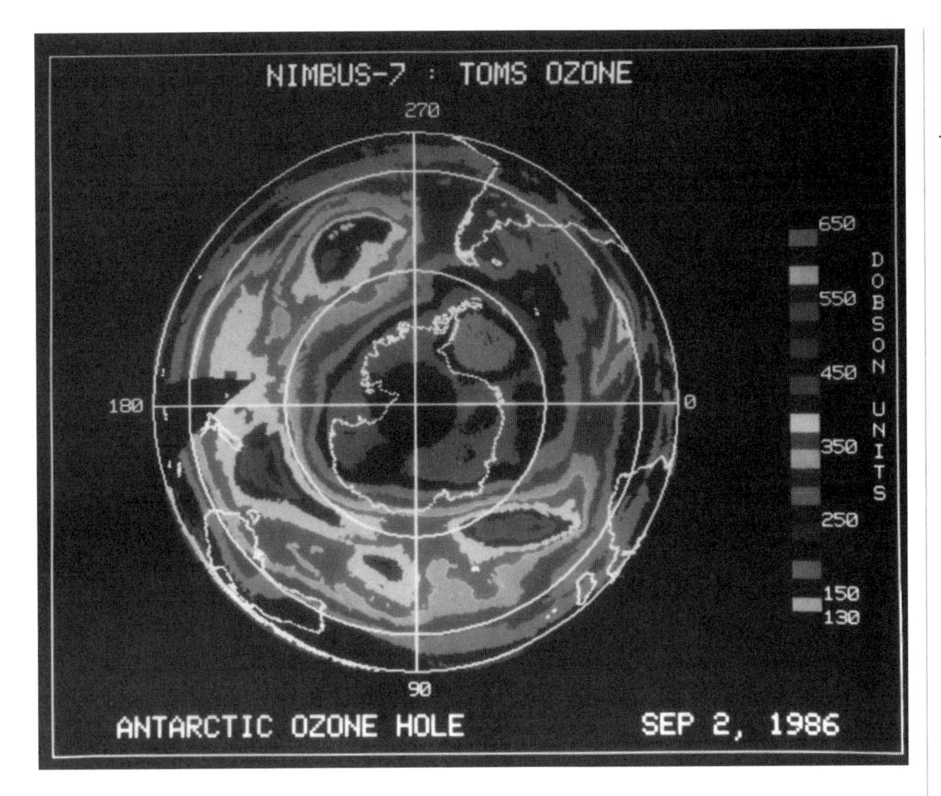

Figure 24 *A map of total ozone in the Southern Hemisphere from* Nimbus 7 *satellite illustrating the Antarctic ozone hole.*

(Photo courtesy NASA)

high ultraviolet exposures. Extreme cold conditions in the polar stratosphere help pollution destroy ozone by chemical processes.

Ozone, a molecule of three oxygen atoms (Fig. 25), constitutes less than one part per million of the gases in the atmosphere. Ozone plays a vital role

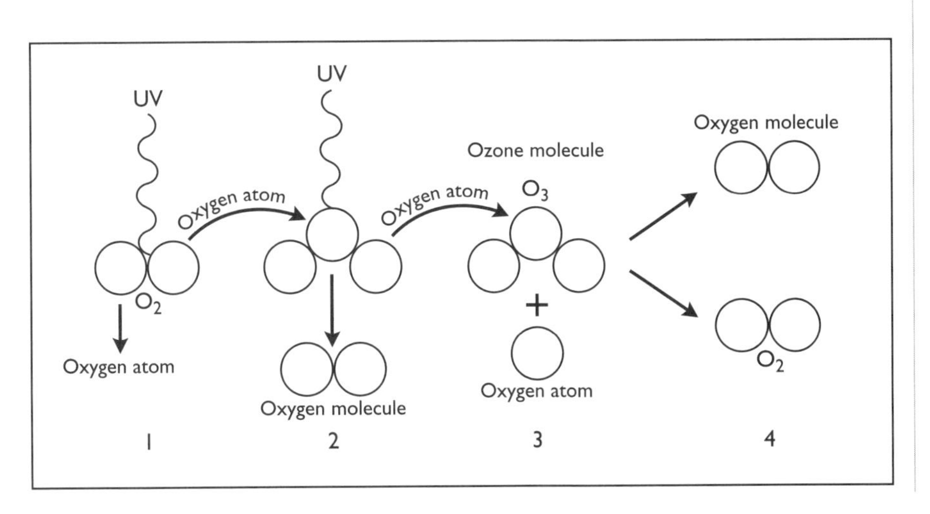

Figure 25 *The life cycle of an ozone molecule.* 1) *Ultraviolet (UV) radiation splits an oxygen molecule into two oxygen atoms.* 2) *One of these atoms combines with another oxygen molecule to create an ozone molecule. The ozone molecule traps UV, liberating an oxygen molecule and an oxygen atom, to reform as an ozone molecule.* 3) *The addition of another oxygen atom creates* 4) *two new oxygen molecules.*

by shielding Earth from the Sun's harmful short-wave ultraviolet radiation, or UV-B. A slight increase in ultraviolet radiation can promote a rise in medical conditions such as skin cancer, cataracts, and weakened immune systems. Ultraviolet exposure can also harm animals and plants, especially crops needed to sustain an ever-growing human population. In addition, high levels of ultraviolet radiation can exacerbate serious pollution problems such as smog and acid rain.

Ozone depletion is strongly believed to be caused by synthetic chemicals, principally halocarbons used as refrigerants and solvents along with nitrogen oxides from the combustion of fossil fuels and deforestation. The threat posed by deforestation is the release of nitrous oxide into the atmosphere, possibly damaging the stratospheric ozone layer. Forest clear-cutting encourages soil bacteria to produce nitrous oxide that escapes into the air. The tremendous heat generated by burning forests combines nitrogen and oxygen to form nitrous oxide. A significant amount of this gas enters the upper atmosphere, where it destroys ozone. Large volcanic eruptions, such as Mount Pinatubo, Philippines, in June 1991, also contribute to ozone depletion by sulfuric acid emissions.

Long-term records show that ozone levels in the high northern latitudes have dropped about 5 percent over the last two decades. They could drop another 5 percent early in this century if trends continue. Each percentage point drop in the ozone level could result in a 2 percent rise in the incidence of skin cancer. Unfortunately, even if the chemical emissions ceased entirely, the ozone layer would continue to diminish for at least another century. This is the time required to cleanse the upper atmosphere of ozone-destroying chemicals, which have half-lives of 50 to 100 years, allowing them enough time to accumulate and reach the stratosphere.

Industrial activities account for injecting roughly 10 times more sulfur into the atmosphere than do natural sources such as volcanoes. The industrial era brought about the combustion of high-sulfur coal and oil along with the smelting of sulfide ores, particularly in the heavily industrialized regions of the Northern Hemisphere. Sulfates, which are the leading constituents of acid precipitation, also cut visibility by upward of 50 percent or more. This leaves many parts of the world living in a constant haze.

Factories, power plants, and motor vehicles around the world send aloft thousands of tons of dangerous chemicals into the air annually. Many of these substances rain out of the atmosphere and contaminate the soil and water, where they can concentrate to toxic levels. These highly acidic particles can alter the pH (acid/base) balance of rivers and lakes and also damage forests.

More than 200 hazardous chemicals are vented into the air. Some industrial plants spew millions of pounds of known carcinogens, or cancer-causing agents, into the atmosphere each year. Although no conclusive evidence concerning the health hazards of these toxic pollutants on the population has been found, concern about the long-term exposure to air contaminants warrants further investigation. Several of these substances are rained out of the atmosphere and contaminate soils, rivers, lakes, and the ocean, where they can become concentrated to lethal levels.

ACID PRECIPITATION

Acid precipitation from the burning of fossil fuels is a growing threat to the environment. Despite pollution controls and much scientific study (Fig. 26), acid rain remains a threat to wildlife habitats. It is especially harmful to aquatic organisms. Most aquatic species cannot tolerate high acid levels in their environments. The damage is due to nitrogen oxides in acid rain. Nitrogen is a nutrient that promotes the growth of floating algae, which blocks sunlight and halts photosynthesis below the water surface. When the algae die and decompose, bacteria deplete the water of its dissolved oxygen, which in turn suffocates other aquatic plants and animals.

The ocean is not immune from pollution, either. It is contaminated from widespread increases in nitrate levels along with higher concentrations of toxic metals, including arsenic, cadmium, and selenium. The main factors contributing to these increases are fertilizer, herbicide, and pesticide runoff along with

Figure 26 *Small, open-top chambers for acid rain study.*

(Photo by Dorothy Andrake, courtesy USDA Forest Service)

acid rain, which dissolves heavy metals in the soil. The pollution of many seas has also caused the decline of fisheries. Fish species are rapidly disappearing throughout the world due to deforestation and acid rain. Deforestation causes increased sedimentation. Acid rain acidifies lakes and streams. Moreover, riverine fisheries have been damaged by increased sedimentation from erosion and deforestation in catchment areas that supply water for rivers and aquifers.

Acid rain is a direct consequence of the self-cleaning nature of the atmosphere (Fig. 27). The combustion of high-sulfur fuels mostly in coal-fired furnaces produces sulfur dioxide gas. In the atmosphere, sulfur dioxide readily reacts with water vapor to form fine particles or aerosols. These scatter sunlight and produce a milky white haze common in many cities. The sulfur dioxide also reacts with oxygen in the atmosphere to yield sulfur trioxide. This combines with atmospheric moisture to produce sulfuric acid. In addition, nitrogen oxides from high-temperature combustion produce nitric acid in much the same manner. The acids mix with cloud moisture, forming an extremely corrosive acid precipitation.

The acidity levels of rain and snow indicate that in many parts of the world, especially in eastern North America and northwestern Europe, precipitation has

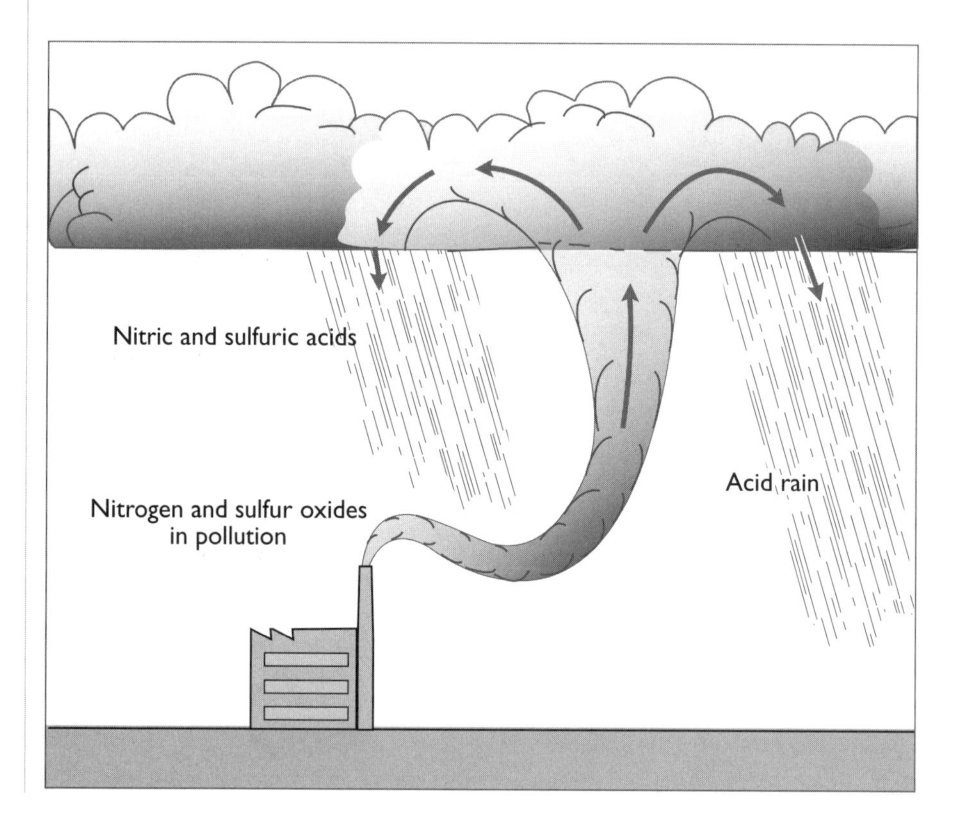

Figure 27 *Acid rain production by cloud scavenging of air pollutants.*

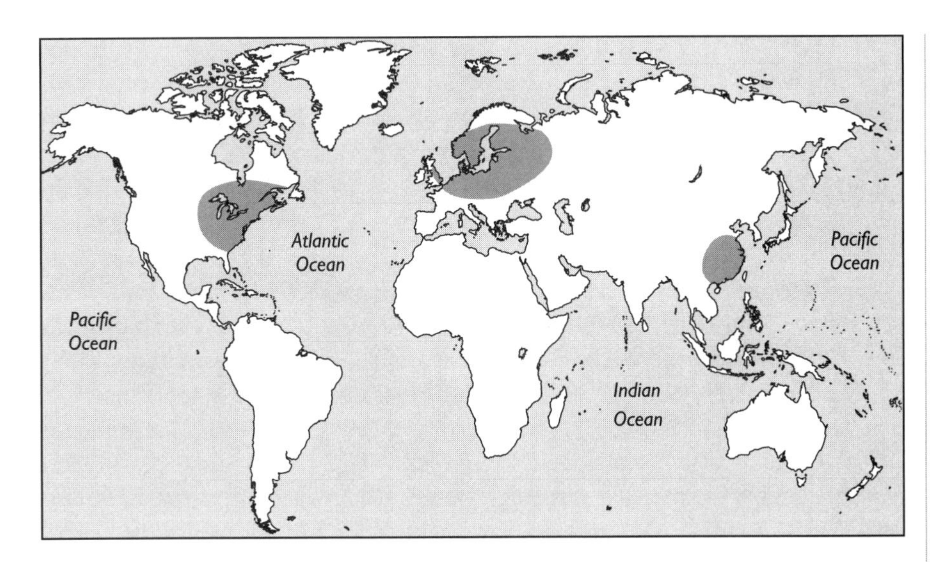

Figure 28 *Areas of heavy acid precipitation in the world.*

changed from nearly neutral at the beginning of the industrial era to a dilute solution of sulfuric and nitric acid today. In the most extreme cases, the rain has the acidity of vinegar. Even in virtually unindustrialized areas such as the Tropics, acid precipitation occurs mostly from the burning of rain forests. Logging and fires open the forest canopy, dry out the forest floor, and increase the risk of massive forest fires. On a global scale, such forest degradation releases large amounts of acid-producing nitrogen oxide as well as carbon dioxide.

High acidity levels in the environment destroy forests, crops, fish, and much of the ancient beauty handed down from earlier civilizations. Acid precipitation, which damages vegetation by harming foliage and root systems, is adversely affecting agricultural crops. It is destroying the great forests of North and South America, Europe, and China (Fig. 28). The acids also deplete soils of valuable nutrients needed by plants for healthy growth. Resorts and wilderness areas, such as those in the western United States, Norway, and Germany, are losing much of their natural beauty due to acid rain. Mountain forests are particularly at risk when covered by acid clouds because usually the cloud base is much more acidic than the acid rain they produce.

Many forests in the Northern Hemisphere are also succumbing to air pollution. The Swiss Alps are threatened with deforestation from air pollution generated largely by heavy automobile traffic. Ironically, the tourist trade the mountains help foster is in the very process of destroying their great beauty. More than half of all alpine trees are sick, and many are dead or dying. The weakened state of these forests leaves them susceptible to disease. This might explain why half the native plants and animals considered endangered or threatened are alpine species.

The temperate forests of the higher latitudes are in danger of destruction as well. Over the past decades, the growth rate and general health of forests in the northeastern United States, eastern Canada, and many parts of central Europe have been declining. Several factors are responsible for destroying the forests along with their water resources. The worst is acid rain. The acids are generated by large industrial centers and precipitated some distance away.

The acidified runoff flows into streams and lakes. It also percolates into the soil, where it damages plant roots, kills nitrogen-fixing bacteria, and leaches out valuable soil nutrients. One important nutrient is calcium, which dissolves in acid rain, leaving the soil calcium deficient. The lack of calcium in their diets has led birds to lay thin eggshells, causing a decline in bird populations. The direct contact of acids on foliage also destroys trees as well as agricultural crops.

Besides acid rain, the added component of acid snow, acid fog, and acid dew are highly destructive. Acid dew forms when dewdrops absorb atmospheric nitric acid gas and sulfur dioxide, which oxidizes to form sulfuric acid. It is also caused by dry deposition of acid particles and gases settling on wet surfaces. Although acid dew does not rival acid rain as an environmental hazard, it can be damaging. Acid dew might significantly harm trees because evaporation concentrates the acids, which could damage leaf surfaces.

Roughly a third of the sulfur dioxide produced in the United States reaches the ground by way of dry deposition. This might be as destructive to the environment as acid precipitation. Sulfates contribute most of the fine-particle mass over much of the eastern part of the country and many other regions. The sulfate particles are often highly acidic, possibly altering the pH balance as much as does acid rain.

SURFACE-WATER POLLUTION

All the freshwater in the world's rivers and lakes represents a small percentage of the total amount of water on the planet. Only a small fraction of the freshwater supply is available for human needs, however, most of which is used for agriculture. Agricultural chemicals such as fertilizers and pesticides are carried off by the drain water, which empties into streams and rivers and then finally flows to the ocean. There, if concentrations are high enough, the chemicals can kill fish and other marine life.

The world's rivers and coastal waters have become the dumping grounds for millions of tons of toxic wastes yearly (Fig. 29). In addition, raw sewage in coastal treatment plants drains directly into the sea due to overflows or equipment failures during heavy downpours. Besides human effluent, which is extremely toxic, other municipal wastes are disposed of in metropolitan sewage systems. The potential of environmental damage and the spread of diseases

Figure 29 *Water pollution in the Cumberland River, Nashville, Tennessee, on May 12, 1970.*

(Photo by William Bram, courtesy USDS Soil Conservation Service)

have resulted in beach closings in many parts of the world. Many beaches are also littered with refuse (Fig. 30). The pollutants in the Mediterranean Sea, where many beaches are deemed unsafe for swimming, will eventually pollute the rest of the world's oceans. The Caribbean, the North Sea, the Gulf of Finland, and other heavily polluted seas are suffering a similar fate.

Although rivers and enclosed or semienclosed seas are more polluted than the open ocean, its waters are also slowly succumbing to pollution. Even the middle of the Pacific Ocean, once thought to be pristine, is polluted with particulate matter. Ocean currents often bring the wastes back to shore. Other wastes are concentrated between thermal layers and ocean fronts, where lie some of the world's most productive fishing grounds.

Some pollutants are powerful carcinogens and mutagens. Many are nonbiodegradable and remain in the environment for extended periods. Toxic substances diluted to supposed safe levels in lakes and streams are concentrated by biologic activity. At the base of the food chain, toxins accumulate in primary producers. These are eaten by fish and other aquatic life, some of which are a major dietary source for humans. Mercury poisoning of fish by industrial wastes dumped into rivers works by just such a process.

Acid runoff along with direct deposition of acid precipitation taints once pristine lakes with levels of mercury sufficiently high to be a public health

Figure 30 *A littered beach on Clear Lake near Natchitoches, Louisiana.*

(Photo by M. J. Hough, courtesy USDS Soil Conservation Service)

risk. Mercury also descends upon lakes as fallout from distant sources such as coal-fired electrical generating plants, smelters, and incinerators. Many lakes in the United States, Canada, and Sweden harbor fish populations with concentrations of mercury well above safe dietary limits. The fish are essentially unaffected by low levels of mercury and can accumulate substantial amounts without serious ill effects. However, higher animals including humans can suffer adverse health effects by eating mercury-tainted fish.

Lakes and streams, especially those not buffered by carbonate rocks, which help neutralize acid, have become so acidic from acid rain runoff or polluted by toxic wastes that fish populations have been nearly decimated (Fig. 31). Acid rain does not just kill fish directly. It also undermines aquatic food chains, altering the organic composition of the lakes. Much of the damage arises from nitrogen oxides in acid rain.

Many rivers, especially in eastern Europe, which has little or no environmental regulations, are in a dismal condition. The southern Volga River in Russia is on the brink of disaster from millions of tons of solid waterwaste dumped into it annually. Furthermore, massive amounts of water are drawn from the river for agriculture and industry. The Aral Sea has dropped some 50 feet in the past quarter century because rivers that feed it have been diverted for irrigation. Because its waters are so polluted, the Aral can no longer be safely fished. The Sea of Asov is suffering a similar fate due to pesticide runoff. The Baltic Sea has been heavily polluted, creating serious problems for Finland, Sweden, and Denmark.

Even the world's most pristine lakes, including Baikal in Russia, the deepest and largest in the world by volume, are being threatened by pollution. The rising, polluted waters of the Great Lakes have threatened beaches and coastal homes. Toxic pollutants rain directly into the lakes or run off contaminated areas onshore. These toxins are the leading environmental issue for the Great Lakes because they require a time scale of 100 years or more to drain the polluted waters into the Atlantic Ocean.

The coastal seas of the world are among the most fragile and sensitive environments. Some changes in the ocean environment are irreversible. One example is damming rivers, which limits the amount of water discharged into the sea. Another is building ports at the mouths of estuaries, which permanently modifies the patterns of water flow and alters the coastal habitat.

Oil spills are by far the most damaging of all coastal pollution (Fig. 32). Hydrocarbon chains called surfactants coat the surface of the ocean with a thin film that interferes with the transfer rates of gas and water vapor between the ocean and the atmosphere. One example is ordinary soap, which is a dry surfactant because it mostly rides on the surface of the water. This type of surfactant is

Figure 31 *Fish killed by the pollution of Frene Creek, Hermann County, Missouri, from an unknown source. The estimated kill was between 10,000 and 15,000 fish.*

(Courtesy USDA Soil Conservation Service)

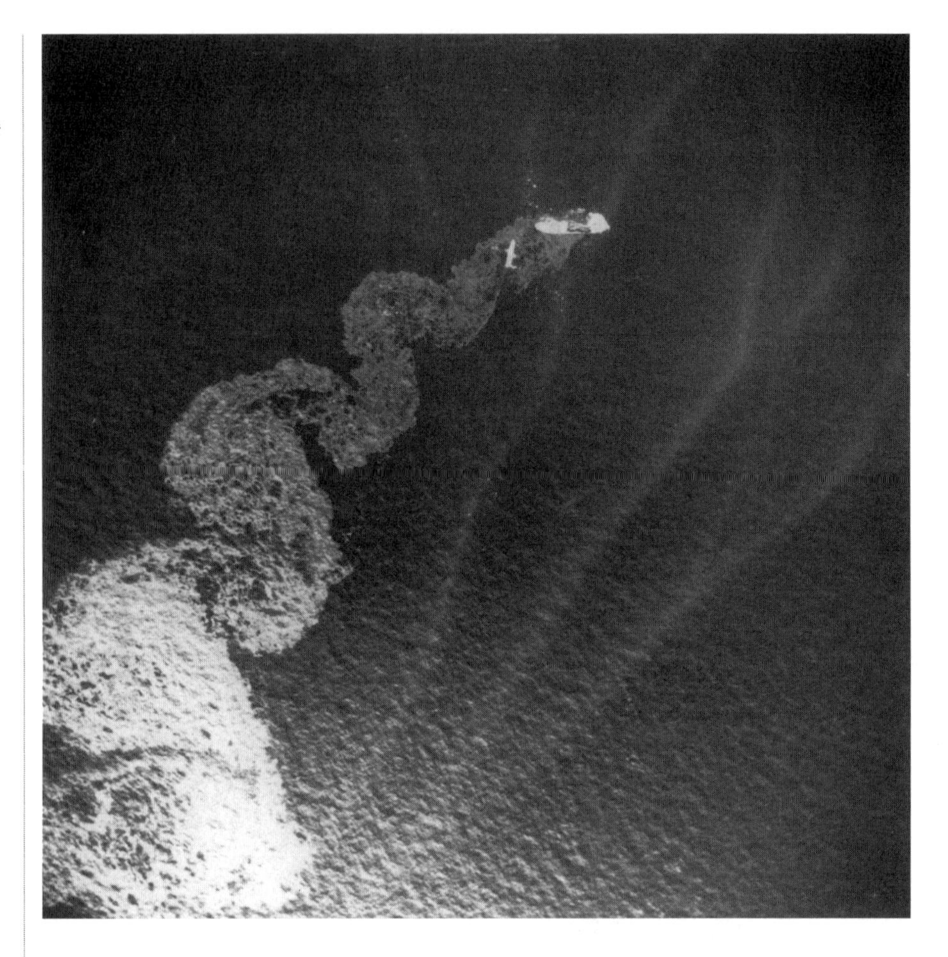

Figure 32 *Aerial view of the December 19, 1976, Argo Merchant oil spill 28 miles off the coast of Nantucket, Massachusetts.*

(Courtesy NASA)

rare except in areas of man-made pollution, especially oil spills, which produce a thin, suffocating film that rides on the surface.

The number of oil spills is increasing steadily as consumption rises. Every year, millions of barrels of oil spill into the world's oceans. Increasing demand for offshore oil, collisions and groundings of oil tankers, attacks on oil tankers by warring nations, and deliberate dumping of oil into the sea as environmental terrorism have led to disastrous ecological consequences.

In the 1991 Persian Gulf War, the Iraqis deliberately dumped more than 1 million barrels of crude oil from five supertankers berthed at the Kuwaiti port of Mina. They also discharged another million or more barrels of oil from a tanker-loading terminal 10 miles offshore. This act of environmental terrorism created the world's largest oil spill. It sent as much as 3 million barrels of crude into the shallow and relatively enclosed waters of the Persian Gulf.

This terrorist act caused unprecedented damage to the ecology of the Persian Gulf for many years. The immediate damage was to seabirds caught in the oil slick washing up onshore. Fish and shellfish also suffocated under the blanket of crude. Some 6 million barrels of crude, amounting to about 10 percent of the world's daily consumption of oil, went up in smoke daily when Kuwaiti oil wells were set on fire. The fires sent 50,000 tons of sulfur dioxide and 100,000 tons of soot into the atmosphere every day. It was considered the worst environmental catastrophe in modern history.

GROUNDWATER POLLUTION

Groundwater pollution has become the primary environmental challenge for the future. Subsurface water, which underlies millions of square miles of the United States, holding an estimated 65 quadrillion gallons of water, is becoming increasingly polluted. Throughout the country, toxic chemicals are leaching out of landfills. Agricultural pesticides and fertilizers are penetrating into the ground. Contaminants are percolating down through layers of soil into groundwater aquifers. Dangerous chemicals seeping into groundwater supplies place many municipalities at risk from contaminated drinking water. Hazardous wastes, including organic chemicals, heavy metals, pesticides, and other toxic substances, seep into the ground from landfills, buried gasoline tanks, septic systems, radioactive waste sites, farms, mines, and many other sources.

The continental movement of subsurface water is extremely slow, taking as long as 1 million years. As much as 10 percent of the groundwater supply across the nation is contaminated. This could become worse as slow-moving patches of pollutants advance on populated areas. An increase in groundwater use also exacerbates the problem because overpumping speeds the flow of water along with contaminants in an aquifer. Contamination of the aquifers is so extensive that in the ensuing years, half the nation's groundwater could be rendered useless.

Water wells across the country have been contaminated by highly concentrated chemicals that spread through aquifers and exceed safe drinking water limits. Waste lagoons and settling ponds contain toxic substances that pollute the groundwater. Solvents and other chemicals used in manufacturing are leaking from buried storage tanks into the water supplies of many communities. Landfills containing hazardous wastes are contaminating aquifers and forcing the shutdown of nearby water wells.

The sources of pollution are so diverse that determining the major cause of groundwater contamination is often difficult. Routine monitoring of industrial waste lagoons and landfills reveal whether the chemicals are being contained or are contaminating nearby water wells. Pools of chemicals generally advance at the rate of the subterranean flow. Once the pollution is located, determining its extent is often difficult and expensive. The cost of cleaning the toxic sites is also phenomenally expensive.

Cleaning up aquifers is prohibitively costly. In the simplest cases when the contamination is fairly localized, the pollutants can be pumped out of water wells. If the chemicals have spread over a wide area, they can be blocked or encircled with impermeable clay pumped into wells drilled into the formation. However, these methods work only when the contamination originates from a single source and covers a limited area. If the contamination is irreversible, the only recourse is to treat the water at the wellhead.

WASTE DISPOSAL

The disposal of wastes generated by modern society remains one of the most pervasive problems for the forthcoming years (Fig. 33). As the number of dump sites dwindles, the garbage continues to mount. Most of the waste is trucked to already overflowing landfills and buried under such conditions they cannot deteriorate properly.

Figure 33 *Trash and debris strewn along the banks of the Brandywine River near Greenfield, Indiana.*

(Photo by E. W. Cole, courtesy USDA Soil Conservation Service)

Figure 34 *Smoke and refuse from garbage incineration cause considerable land, air, and water pollution near Brunswick, Maine.*

(Photo by Richard Duncan, courtesy USDA Soil Conservation Service)

Landfills, in the short term, will probably continue to receive most of the trash largely because they are inexpensive compared with other methods of dealing with garbage disposal. However, in most major cities, landfills have reached capacity. Few new spaces are available to put the trash. Often, toxic substances leach out of landfills and contaminate nearby water wells, requiring expensive treatment. Many toxic pollutants are powerful carcinogens and mutagens. Some are nonbiodegradable and persist in the environment for long periods.

Of all the alternative methods for the disposal of waste materials, none has engendered more controversy than incineration (Fig. 34). Unfortunately, with incineration, the garbage disposal problem is solved at the expense of creating an enormous air pollution problem. Thousands of tons of pollutants, including toxic dioxins and other dangerous chemicals, would be emitted into the air each year. Even open burning of trash creates a serious pollution problem. Every 100 tons of trash incinerated generates 30 tons of ash, often laden with heavy metals. This qualifies as a hazardous waste, creating another garbage disposal problem.

The dumping of toxic wastes into the ocean is an insidious and potentially serious problem. Because of the escalating cost of land disposal of municipal and industrial wastes, many coastal metropolitan areas around the world are forced to dump them directly into the sea. Much of the waste that washes to shore originates from overburdened sewage-treatment plants, accidental spills by garbage barges, and absence of winds and ocean currents to disperse the flotsam. Millions of tons of toxic wastes are dumped into rivers and coastal waters each year. Some of these toxic pollutants are powerful carcinogens and mutagens. Many are nonbiodegradable and persist in the environment for extremely long periods.

The disposal of radioactive wastes from nuclear power plants, hospital radiation labs, and nuclear weapons manufacturing (Fig. 35) has received much attention in recent years. Concerns are increasing over the long-term environmental effects and because of the expansion of nuclear technology throughout the world. As the demand for nuclear-generated electricity rises, a viable solution for the storage of nuclear wastes must soon be found. High-level nuclear wastes are the most difficult radioactive waste materials to dispose of because of their high radiation and heat output and their longevity. Some substances such as plutonium require millions of years to decay.

The best place to store nuclear waste is thought to be deep underground (Fig. 36). Much effort has been focused on exploring for stable geologic formations free from earthquakes and volcanic eruptions. Eruptions are of particular concern because they could blast radioactive material into the atmosphere. Earthquakes could open fissures in the crust to allow the escape

Figure 35 *Rocky Flats nuclear materials production plant, Golden, Colorado, in 1988.*

(Photo courtesy U.S. Department of Energy)

Figure 36 An underground nuclear waste disposal site in a salt bed 2,000 feet below the surface near Carlsbad, New Mexico.

(Photo courtesy U.S. Department of Energy)

of nuclear wastes. Salt domes and granite make the most stable geologic formations on the continents. However, mine repositories are very expensive and require additional costs for backfilling and shaft sealing. Once buried, the waste must remain isolated for thousands of years while its radioactive isotopes decay to prevent theft of nuclear-bomb-making materials.

After the nuclear waste containers age and begin to leak, they must not contaminate nearby groundwater systems. Therefore, reliable predictions concerning the possible migration of radioactive fluids through geologic formations surrounding the repository must be made. The formation must remain stable for 1 million years or so without earthquakes or other geologic activity. It must also be guarded against intrusion and theft for countless generations to come.

Transporting the nuclear wastes to dump sites in the West is also hazardous, especially since 85 percent of the waste is scattered in various parts of the nation, mostly in the East. The U.S. Department of Energy estimates that 17 truckloads per day for the next 20 years would be required to move all the wastes to the burial sites. With that many vehicles carrying nuclear wastes on American roads at any given time, the prospect of an accident and a consequent spill is far too great for communities along the routes to accept. Packaging the nuclear wastes in solid form or designing containers that are virtually indestructible might alleviate the danger from highway accidents.

Other proposals call for building hazardous-waste and nuclear-waste disposal plants on Indian reservations. One reason garbage companies are looking at Indian lands is that environmental regulations are less rigid there than elsewhere. Many Indian tribes, however, refuse to poison their sacred lands.

Another suggestion is putting the nuclear wastes in wells drilled deep into the seabed. The idea is that certain parts of the ocean floor are the most stable environments on Earth, whereas the land is always subject to volcanoes, earthquakes, mountain building, erosion, and leakage into the groundwater system. Once the nuclear waste containers are sealed against the sea, the constant rain of detrital material washed off the continents will continue to bury them under thick layers of sediment. Areas containing natural resources such as fisheries, petroleum reserves, or mineral deposits must be avoided for fear of disturbing the burial site and contaminating the ocean.

Nuclear accidents, such as the 1979 radioactive spill at the Three Mile Island nuclear-generating plant (Fig. 37) and the 1986 reactor explosion at Chernobyl, Ukraine, reveal the dangers of nuclear fission energy. As human populations continue to grow and the demand for nuclear-generated electricity continues to rise, presently accounting for about 15 percent of the

Figure 37 *The Three Mile Island nuclear power plant, Harrisburg, Pennsylvania. It underwent a reactor accident in 1979, which necessitated reassessment of nuclear safety in the United States.*

(Photo courtesy U.S. Department of Energy)

world's total generating capacity, a viable solution for the storage of nuclear wastes must soon be found if Earth is to remain free of radioactive poisons.

ENVIRONMENTAL RESTORATION

The atmosphere's cleansing agent is a short-lived molecule of one hydrogen atom and one oxygen atom called hydroxyl, which strongly reacts with most atmospheric contaminants, rendering them harmless. Unfortunately, with the mounting load of atmospheric pollutants caused by human activities, mostly carbon monoxide and methane, the amount of hydroxyl is declining just when it is needed the most. Since the beginning of the industrial era, hydroxyl has decreased by as much as 20 percent. Further declines could mean much dirtier skies in the future.

Noctilucent clouds, which streak the high polar atmosphere—called the mesosphere—just after dark on summer nights, have been growing steadily brighter and becoming more common than usual. Apparently, the mesosphere is starting to feel the far-reaching effects of global warming from atmospheric pollution. Atmospheric methane, which breaks down by sunlight into water molecules in the mesosphere, has doubled since 1900. Additionally, noctilucent clouds have grown nearly 10 times brighter.

Hazardous wastes, including synthetic organic chemicals, heavy metals, pesticides, and other toxic substances, seep into the ground from landfills, buried gasoline tanks, septic systems, radioactive waste sites, farms, and mines. Toxic chemicals percolating into the ground contaminate aquifers, requiring the testing of nearby water wells. In many cases, the cleanup of aquifers is nearly impossible. If the contamination is irreversible, the water must be treated at the wellhead at great cost. Most states that bear the burden of monitoring and safeguarding the groundwater cannot afford the enormous expenditure without federal government support.

The cleanup of landfills and waste lagoons is also expensive. Bioremediation, which employs microbes to digest toxic wastes, can be used to clean up soil contaminated by underground storage tanks. However, the only remedial action for soils contaminated with toxic or radioactive wastes is to transport them to hazardous-waste dumps. Ultimately, improved methods of dealing with waste problems and better knowledge of the underground environment can help solve future pollution problems. Unfortunately, for much of the nation's groundwater supply, past mistakes might make recovery too late.

Combating oil spills on the ocean requires containment, using floating booms or an absorbent material such as straw. Burning the oil eliminates most of the pollution on site soon after the spill before it spreads or coagulates with seawater. If the oil washes to shore, a labor-intensive cleanup is required (Fig. 38).

Detergents or other chemicals can break up and disburse the oil. However, many of these chemicals are toxic and can cause additional ecological damage. Cleaning up oil-soaked beaches requires chemical dispersants or steam, which kill organisms that otherwise would have survived if left alone. Often, over time, nature does a better job of cleaning up oil-soaked beaches than people do.

To eliminate local air pollution from coal-fired plants, companies build towering smokestacks (Fig. 39) to mix the pollutants with the turbulent air above, where they are carried away by the wind. Acid precipitation known to exist for many decades near large cities and industrial plants prompted the construction of tall smokestacks to disburse emissions high into the atmosphere and away from cities. Unfortunately, this solution only turns a local pollution problem into a regional one as pollutants travel long distances, even crossing international boundaries, causing political problems downwind. For example, Canada suffers from pollution from the industrialized areas of the Ohio River Valley of the United States. Sweden and Norway, normally pollution free by themselves, are constantly being bombarded by pollution from the heavily industrialized regions of Europe.

Yet despite cuts in power plant emissions in the United States, these actions have not significantly reduced acid rain damage in the Northeast. Lakes, streams, soil, and vegetation continue to suffer even after the emissions reductions mandated by the Clean Air Act. Only deeper reductions in oxides of nitrogen and sulfur, the pollutants that cause acid rain, are likely to help regions recover from decades of damage. Coal-burning plants in the Ohio

Figure 38 *A Forest Service biologist checks an area that was cleaned by Exxon workers using high-pressure hot water, following the 1989 Alaskan oil spill.*

(Photo by Jill Bauermeister, courtesy U.S. Forest Service)

Figure 39 *Clean coal technology demonstrated at the Yates electrical generation station, Coweta County, Georgia.*

(Photo courtesy U.S. Department of Energy)

River Valley are a major source of nitrogen and sulfur pollution that contaminates the Northeast. The pollution travels eastward on prevailing winds, mixes with atmospheric moisture, and precipitates as acid rain, snow, or fog.

Acid precipitation can be reduced by installing scrubbers on coal-fired plant smokestacks and by using low-sulfur coal to eliminate sulfur dioxide. Many coal-fired electrical generating plants could convert to natural gas, which is a much cleaner burning fuel, producing only half the amount of carbon as does coal per unit of energy. However, rapid rises in natural gas prices have made clean coal-burning technology a more desirable alternative. Many older plants built before 1975 and plants built in other countries are not required to make costly investments to clean up the air, however. Governments are generally reluctant to pass laws requiring mandatory emission controls to clean up the environment because of possible adverse economic effects.

Dumping carbon dioxide underground or into the oceans could slow global warming. Plankton might be encouraged to grow more vigorously in the ocean so the organisms could absorb more nutrients and enhance photosynthesis. They, in turn, would absorb more carbon dioxide from the atmosphere.

One idea is to seed the ocean with iron. Presently, the main source of iron in the ocean is dust blown off the world's deserts. But a single supertanker load of iron could trigger enough phytoplankton growth to draw 2 billion tons of carbon out of the atmosphere. However, upward of a century or more of iron fertilization might be required to reduce atmospheric carbon dioxide levels significantly.

Many lakes and streams in the United States have higher than normal acidity levels. Sensitive aquatic species can die from long-term exposure to low acidity levels or short-term exposure to high acidity levels. High acidity levels in lakes and streams can be suppressed by treatment with lime made from abundant limestone, which neutralizes the acids. Treating the watersheds that feed the streams and lakes is more effective than applying the lime directly. Furthermore, halting acid precipitation at the source by treating the flue gases at coal-fired plants with lime could clear up half the acidic surface waters.

The environment would require years or even decades before fully recovering from acid precipitation. Despite tighter controls on pollution, rain falling onto parts of North America and Europe are still strongly acidic. The reason is that levels of atmospheric dust particles that neutralize acids in the air are dropping because the emissions of base elements such as calcium and magnesium have fallen. These elements are also essential nutrients for most plants, and soil depletion could lead to poor productivity.

After discussing pollution and its effects on the environment, the next chapter investigates how atmospheric pollutants affect the climate.

3

CLIMATE CHANGE

THE GREENHOUSE EFFECT

This chapter examines how greenhouse gases and other contaminants in the air affect the climate. Human activities and natural phenomena, including volcanic eruptions, are responsible for changing the gaseous composition of the atmosphere. They threaten to produce global climate change. Increasing amounts of atmospheric carbon dioxide along with other greenhouse gases, principally methane, tend to warm the planet and energize the atmosphere. This, in turn, alters the hydrologic cycle, responsible for bringing life-sustaining rains to all parts of the world.

The rise in global temperatures could also cause the melting of the polar ice. This would raise sea levels and flood coastal regions, where half world's population lives. Shifting precipitation patterns could cause serious drought and desertification in some regions and severe flooding in others. Changing atmospheric circulation patterns could significantly affect the weather, producing violent storms and potentially much death and destruction.

CLIMATES FOR HUMANITY

The climate has always been changing. Several hundred million years ago, the global environment was quite different from today. The climate of those times would have been quite inhospitable to human beings. The human species arose some 4 million years ago during a period of relatively stable climate, when species diversity was at an all-time high.

During a warm interglacial period about 400,000 years ago, the climate was much warmer than today. Sea levels were 60 feet higher due to the melting of the ice caps. In the last ice age, beginning about 125,000 years ago, lowered precipitation levels caused desert regions to expand in many parts of the world. Desert winds blew with such a force they generated gigantic dust storms that blocked out sunlight and cooled the planet. When the great ice sheets retreated to the poles, tropical regions of Africa and Arabia began to dry out during a period of rapid warming. The climate change resulted in the expansion of the arid regions between 14,000 and 12,500 years ago.

During a wet period, from 12,000 to 6,000 years ago, some of today's African deserts were covered with lush vegetation and contained many large lakes. Lake Chad on the southern border of the Sahara Desert was Africa's largest lake at 10 times its present size. Lakes in other parts of the world were similarly affected. Utah's Great Salt Lake occupied the adjacent salt flats (Fig. 40), expanding several times its current size.

One of the most dramatic climate changes in the history of the planet occurred during the present interglacial, called the Holocene epoch in geology. It is also known as the Neolithic in archaeology and is commensurate with the rise of civilization. The climate 10,000 years ago at the beginning of the Holocene was significantly different from that of the previous 10,000 years during the height of the last ice age. During the last 8,000 years, Earth's climate has been extremely beneficial. People have prospered exceedingly well under these benign conditions.

Following the retreating glaciers of the last ice age, plants and animals returned to the northern latitudes. A long wet spell during this time might have been caused by the strengthening of the monsoons, which carry moisture-laden sea breezes inland over Africa, India, and Southeast Asia. The continental interiors 9,000 years ago were warmer in summer, which invigorated the monsoon winds. The Climatic Optimum, which began about 6,000 years ago, was a period of unusually warm, wet conditions that lasted for 2,000 years. During that time, early civilizations prospered well.

A relatively mild arid episode between 7,000 and 6,000 years ago was followed by a severe 400-year-long drought starting about 4,000 years ago. Temperatures dropped significantly. The world became drier, forming today's deserts (Fig. 41). Around 1,000 years ago, the world warmed again, during the

Figure 40 *The south end of Great Salt Lake Desert from southwest of Simpson Range, Tooele County, Utah in 1903.* (Photo by C. D. Walcott, courtesy USGS)

Medieval Climate Maximum. Some 500 years later, the world plunged into the Little Ice Age, which lasted three centuries while average global temperatures dropped about 1 degree Celsius.

The expanding glaciers forced people out of the northlands of Europe and decimated the Greenland Normans, who had successfully inhabited the island for nearly 500 years. By the middle 19th century, the world warmed again until 1938, cooled until 1976, and presently appears to be in another long-term warming trend that shows no signs of abating. As such, the climate appears to be changing again—possibly to one much less to people's liking.

GREENHOUSE GASES

The mechanics of the greenhouse effect was first recognized in 1896, when the Swedish chemist Svante Arrhenius predicted the influence of atmospheric carbon dioxide on the climate. He suggested that past glacial periods might have occurred largely because of a reduction of carbon dioxide in the atmosphere. Arrhenius also estimated that doubling the concentration of atmospheric carbon dioxide would cause a global warming of about 5 degrees Celsius, surprisingly concurrent with present-day greenhouse models.

Figure 41 *Eastern Mojave Desert, San Bernardino County, California.*

(Photo by R. E. Wallace, courtesy USGS)

The loss of greenhouse gases, principally carbon dioxide, by photosynthesis when single-celled plants first evolved might have cooled the climate sufficiently to produce the first known ice age in geologic history between 2.2 and 2.4 billion years ago. The burial of large amounts of carbon in the crust might have been the key to the onset of perhaps the greatest of all ice ages during the late Precambrian era about 680 million years ago. The ice cover was so extensive that the period has been dubbed the "snowball Earth," a time when even the Tropics froze over. If not for massive volcanic activity, which restored the carbon dioxide continent of the atmosphere, the planet would have remained buried under ice.

The glaciations of the late Ordovician around 440 million years ago, the middle Carboniferous around 330 million years ago, and the Permo-Carboniferous around 290 million years ago might have been influenced by a reduction of atmospheric carbon dioxide to about one-quarter of its present value. Another glacial episode about 270 million years ago might have been triggered by the spread of forests across the land as plants adapted to living and reproducing out of the sea. Earth began to cool as the forests removed atmospheric carbon dioxide, converting the carbon into organic matter that formed massive deposits of coal. The conversion to coal buried substantial amounts of carbon in the crust, thereby cooling the climate.

Such a change in carbon dioxide concentrations in the atmosphere might have led to the ice ages. Data taken from deep-sea cores established that carbon dioxide variations preceded changes in the extent of the more recent glaciations (Fig. 42). Possibly, the earlier glacial epochs might have been similarly affected. The variation of carbon dioxide levels might not be the sole cause of glaciation. When combined with other processes, such as variations in Earth's orbital motions, they could be a powerful influence. This might explain why the ice ages have turned on and off again throughout geologic history.

The concentration of carbon dioxide in the atmosphere has increased from 265 parts per million (ppm) in preindustrial times to more than 365 ppm today. Over the last century, surface temperatures have risen 1 degree Celsius. Paradoxically, the lowermost atmosphere, directly responsible for the greenhouse effect,

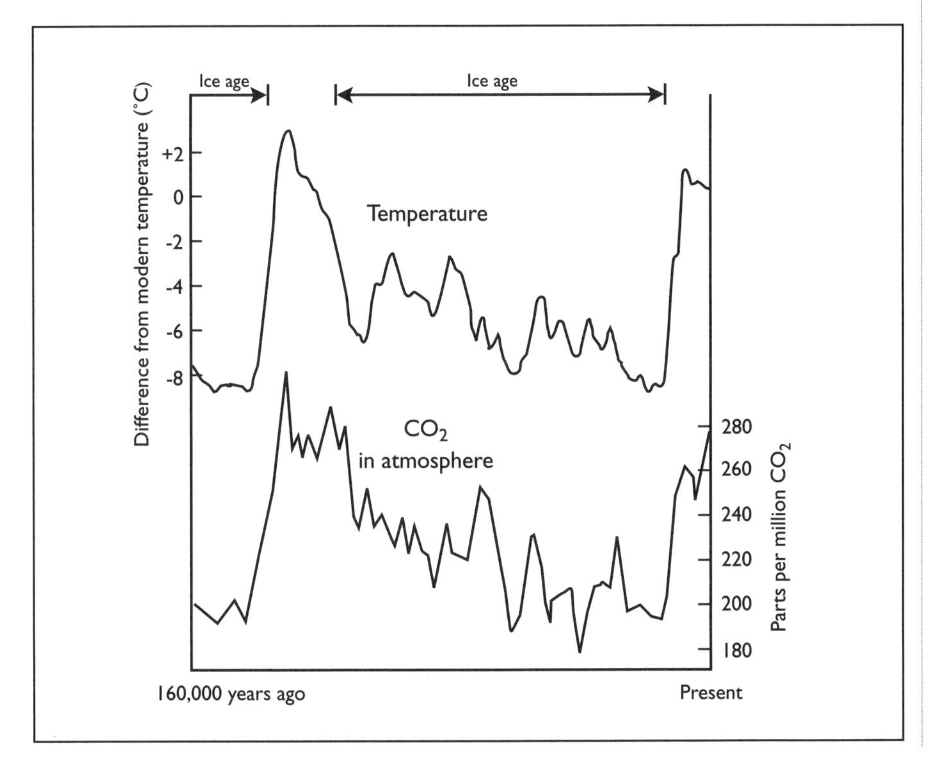

Figure 42 *Global temperatures and atmospheric carbon dioxide levels have kept in step for the past 160,000 years.*

has not warmed nearly as much as measured by satellites and weather balloons. Over the last 40 years, surface temperatures climbed at a rate of about 0.15 degrees per decade, while the atmosphere warmed by only 0.10 degrees per decade.

The atmospheric carbon dioxide content also fluctuates with the seasons, rising to a peak in late winter and falling to a minimum at the end of summer. This is because plants draw carbon dioxide out of the atmosphere and store it in their tissues as carbohydrates during the growing season. The world's great forests have a pronounced influence on the carbon dioxide content of the atmosphere. Much of the seasonal variation in the atmospheric concentration of carbon dioxide is correlated with a rapid rise of photosynthesis during the summer. Forests are extensive and conduct more photosynthesis worldwide than any other form of vegetation. The stores of carbon in the forests are in quantities large enough to affect the carbon dioxide content of the atmosphere substantially.

Forests incorporate from 10 to 20 times more carbon per acre than cropland or pasture, and they contain as much carbon as does the entire atmosphere. The world's forests store a great deal of carbon. The clearing of forest land for agriculture, especially in the Tropics, is a significant source of carbon discharged into the atmosphere. Deforestation accounts for up to one-third the total amount of carbon dioxide and up to half the methane released into the atmosphere. As the carbon stored in the trees escapes into the air, the concurrent reduction of the forests weakens the trees' ability to remove excess atmospheric carbon dioxide, with the potential of causing global warming.

Although the forests of North America and Europe have a net accumulation of carbon, the absorption of excess carbon dioxide is insignificant compared with the losses in the tropical regions. However, the uptake of carbon dioxide has ceased because of massive increases in tree dieback and logging in the northern boreal forests of North America and Eurasia. The increased carbon dioxide content in the atmosphere heightens the greenhouse effect, which could substantially alter global weather patterns.

Carbon dioxide emissions account for almost 60 percent of the annual human contribution of greenhouse gases. Most of the carbon dioxide is generated by industrial nations. Carbon dioxide is responsible for half the greenhouse warming, with water vapor and methane providing the bulk of the rest. Human-generated carbon dioxide also remains in the atmosphere for a century or more. The long-term increase in atmospheric carbon dioxide, as much as 25 percent since 1860, is the result of an accelerated release of carbon dioxide by the combustion of fossil fuels. Each year, human activities emit into the atmosphere 27 billion tons of carbon dioxide, containing 7.4 billion tons of carbon. The present consumption of fossil fuels yields on average about 1 ton of atmospheric carbon for every person on Earth yearly. Americans release some 6 tons per person per year, or about one-quarter of the world's total.

The atmosphere presently holds about 800 billion tons of carbon. Therefore, humans alone are increasing atmospheric carbon by nearly 1 percent annually. Some carbon is removed from the atmosphere by biologic, hydrologic, and geologic processes so that the average annual increase of atmospheric carbon dioxide by human activity is reduced to about 0.5 percent or about 4 billion tons. One-third the carbon dioxide released into the atmosphere originates from the destruction of tropical rain forests and the extension of agriculture mostly by developing countries. These nations will account for a major fraction of greenhouse gas pollution as they attempt to improve their standards of living.

The biota on the surface of Earth and humus in the soil hold 40 times more carbon than the entire atmosphere. The harvest of forests, the extension of agriculture, and the destruction of wetlands speed the decay of humus, which is transformed into carbon dioxide and released into the atmosphere. The destruction of forests alone is responsible for injecting some 2.5 billion tons of carbon into the atmosphere annually. Moreover, agricultural lands, which also produce carbon dioxide during cultivation, do not store nearly as much carbon as the forests they replace. The clearing of land for agriculture, especially in the Tropics, is the largest source of carbon released into the atmosphere by the biota and soils when they are cultivated, which turns over the soil and exposes organic matter to the atmosphere.

Methane is the second most important greenhouse gas. Although methane remains in the atmosphere for only a decade or so, in the presence of oxygen, it eventually breaks down into carbon dioxide. The atmosphere presently contains about one molecule of methane for every 200 molecules of carbon dioxide. However, methane is 20 to 30 times more effective per molecule at absorbing infrared radiation than carbon dioxide. Therefore, even small amounts released into the atmosphere can have a large affect.

Methane production is outstripping that of carbon dioxide and is increasing about 1 percent per year compared with about 0.5 percent for carbon dioxide. In the ensuing years, methane and other trace greenhouse gases such as nitrous oxide generated by agriculture and industry might together contribute more to greenhouse warming than carbon dioxide alone. Nitrous oxide promotes excessive greenhouse warming and has an atmospheric lifetime of more than a century. In addition, every one of its molecules absorbs roughly 200 times more outgoing infrared radiation than does a single carbon dioxide molecule.

Much of the methane production comes from plants and animals. Deforestation is rapidly increasing the number of methane-producing termites due to the large numbers of dead trees. Presently, about three-quarters of a ton of termites exists for every person on Earth. As deforestation escalates, that number could increase significantly. Termites ingest as much as two-thirds of all the carbon available on the land, about 1 percent of which is converted into methane.

Large numbers of cattle contribute substantial amounts of atmospheric methane during their digestive process. From 5 to 9 percent of what a cow eats converts into methane, representing about 5 percent of the total. With one bovine for every four people on Earth, cattle could play a significant role in changing the world's climate. Rice cultivation is another major contributor to the atmospheric load of methane. Natural gas, which is mostly methane, leaks from distribution systems and landfills. A combination of methane and water ice forms a clathrate (a cell-like chemical mixture) in the bottom muds of the deep sea. Ocean warming could expel massive quantities of methane into the atmosphere. Along with carbon dioxide, this could create a vicious cycle of heat and release.

GLOBAL WARMING

Over the last two decades, record-breaking weather events have occurred worldwide. In the United States, the decade of the 1990s witnessed some of the hottest years since the end of the Little Ice Age 150 years ago, even surpassing the Dust Bowl years of the 1930s. These events appear to be symptoms of a global climate change from the chemical pollution of the atmosphere. The strange weather might be just a reflection of natural climate variability. So far, no clear sign of climate change has been found that can be positively blamed on greenhouse warming. The climate changes thought to relate to human activities have been relatively modest. Nevertheless, the degree of change could become dramatic by the middle of this century, possibly exceeding anything seen in nature during the past 10,000 years.

The results of a steady rise in atmospheric carbon dioxide would probably be catastrophic if other moderating factors did not come into play to cancel part of the greenhouse effect. These might include the absorption of excess carbon dioxide and heat by the oceans and green vegetation on land. Most of the carbon that flows through the ocean is organic debris produced by plants and animals. However, these processes act largely to transfer carbon from shallow waters to the abyss and do little to draw carbon dioxide out of the atmosphere. Furthermore, warmer sea temperatures reduce the ocean's ability to absorb excess carbon dioxide and could even increase the expulsion of the gas, similar to a warm bottle of soda.

Where all the carbon dioxide produced by industrial activities is going remains a mystery. Apparently only about 40 percent of the carbon dioxide generated by the combustion of fossil fuels and the destruction of forests is accumulating in the atmosphere and absorbed by the ocean. Some excess carbon dioxide might be taken up by terrestrial vegetation, acting like a fertilizer to stimulate plant growth. Nevertheless, land plants do not store as much carbon dioxide as the oceans and might reach full capacity in the foreseeable future. Furthermore, the continuing destruction of the world's forests dramatically reduces their ability to absorb excess carbon dioxide.

Forests significantly affect the global climate. Deforestation increases surface albedo, allowing more sunlight to reflect into space. Thus, deforestation could cause global cooling that might counteract greenhouse warming by the pollution of the atmosphere with greenhouse gasses. The loss of solar energy from deforestation could change precipitation patterns, causing decreased rainfall, especially in the rain forests themselves. These drought conditions could further stress the trees, making them more susceptible to disease.

Possibly within 50 to 100 years, the world could become hotter than it was 3 million years ago, before the onset of Pleistocene glaciation. The greatest rise in temperature would occur at the higher latitudes of the Northern Hemisphere, with the largest increases during winter. One horrifying aspect of global warming would be the thawing of the Arctic tundra from northern Alaska (Fig. 43) to northern Eurasia. This would release into the atmosphere vast quantities of carbon dioxide and methane trapped in the soil, possibly causing a runaway greenhouse effect. Evidence collected from Alaska's northern tundra suggests that global warming might already be spurring the release of carbon dioxide from the land. The temperature of the Alaskan permafrost is 4 degrees Celsius higher than during the previous years.

The present warming trend amounts to an increase of about 1 degree Celsius over the last 100 years. Since 1850, when the world thawed out following the Little Ice Age, global temperatures have steadily increased (Fig. 44). The rising temperature trend was briefly interrupted by a cooling spell between 1940 and 1976 due to an increase in volcanic activity. Since then, temperatures have resumed their upward rise. Meanwhile, the carbon dioxide content of the atmosphere has increased more than 20 percent over the last century and could double by the middle of this century, possibly raising average global temperatures 1.5 to 4.5 degrees Celsius. Unless nations drastically curtail their use of fossil fuels, carbon dioxide will continue to accumulate even after doubling.

The most unusual aspect of the present global warming trend is its unprecedented speed. The temperature rise is 10 to 20 times faster than the average rate of warming following the last ice age. Between 14,000 and 10,000 years ago, when massive ice sheets across North America and Eurasia melted, Earth warmed perhaps 3 to 5 degrees Celsius. This is comparable to the predicted increase for global warming. The major difference, however, is

Figure 43 *Arctic coastal plain east of Kukpowruk River, Colville district, northern Alaska.*

(Photo by R. M. Chapman, courtesy USGS)

that the temperature rise extended over a period of several thousand years—not compressed into a mere century.

By the end of this century, global temperatures could equal those of 100 million years ago, creating the warmest temperatures since the days of the dinosaurs. Some areas in the Northern Hemisphere would dry out, providing a large potential for massive forest fires. This has dire implications. If greenhouse warming continues, major forest fires, such as those that devastated half of Yellowstone National Park in 1988 (Fig. 45), might become more frequent, with substantial losses of forests and wildlife habitats.

In the past, climate changes occurred slowly enough for the biological world to adapt. However, today's climate changes are much too abrupt, perhaps rapid enough to cause the extinction of plants and animals. Plants would be hardest hit by global warming because they are directly affected by changes in temperature and rainfall. Forests, especially game preserves, might become

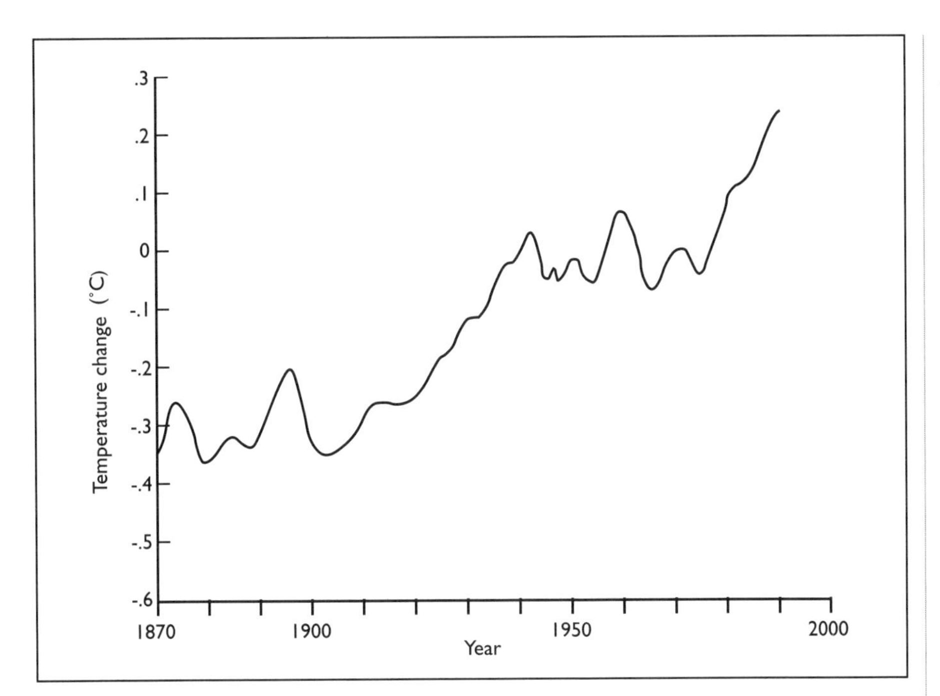

Figure 44 *The rise in global temperatures.*

Figure 45 *A forest fire that engulfed half of Yellowstone National Park, Wyoming, during the summer and fall of 1988.*

(Photo courtesy National Park Service)

isolated from their normal climate regimes, whose climatic conditions would continue to advance to higher latitudes.

The effects of global climate change with increasing temperatures and rainfall would be felt for centuries. During that time, forests would creep poleward, while other wildlife habitats, including the Arctic tundra, would disappear. Many species would be unable to keep pace with these rapid climate changes. Those able to migrate could find their routes blocked by natural and man-made barriers, such as cities and farms. The warming would rearrange entire biological communities and cause many species to become extinct. Others commonly called pests would overrun the landscape.

If global warming trends continue, by the middle of this century, the forests of the southern states could disappear and be replaced by grasslands. The great forests of eastern North America could shift northward as far as 600 miles. Pine forests could take over stands of spruce in New England and the Appalachian Mountains. Pines could reach as far north as the Arctic tundra. The prairies of the Midwest could sweep eastward as far as Pennsylvania and New York. Higher concentrations of atmospheric carbon dioxide, which functions as a fertilizer, favors the growth of weeds. The warmer climate would be a boon for parasites and pathogens, including bacteria and viruses. It could cause an influx of tropical diseases into the Temperate Zone. Disease-carrying mosquitoes that normally die out in cold winters could invade formerly forbidden territories, infecting many more people.

If global temperatures rise too rapidly, forests might fail to keep up with the movement of climate zones toward higher latitudes. This would cause a further decline of the world's forested regions along with the species they support. The resulting ecological change could take several centuries to stabilize, rivaling the environmental conditions that existed at the end of the last ice age some 12,000 years ago.

Increasing surface temperatures resulting from doubling the amount of atmospheric carbon dioxide could adversely affect global precipitation patterns. This would especially occur during El Niño, the warm inshore current flowing from Central America, which upsets the weather around the world (Fig. 46). The enhanced frequency of El Niño events might also be a symptom of greenhouse gas pollution and global warming. During El Niño, the oceans absorb more carbon dioxide than usual. However, the land surface emits more of the gas. These tend to balance each other. La Niña, a periodic drop in sea-surface temperatures in the equatorial Pacific, produces the opposite effect of El Niño, accompanied by above-average precipitation in many parts of the world.

Subtropical regions, between 20 and 50 degrees north latitude and 10 to 30 degrees south latitude, might experience a marked decrease in precipitation, encouraging the spread of deserts. Increasing the area of desert and semidesert

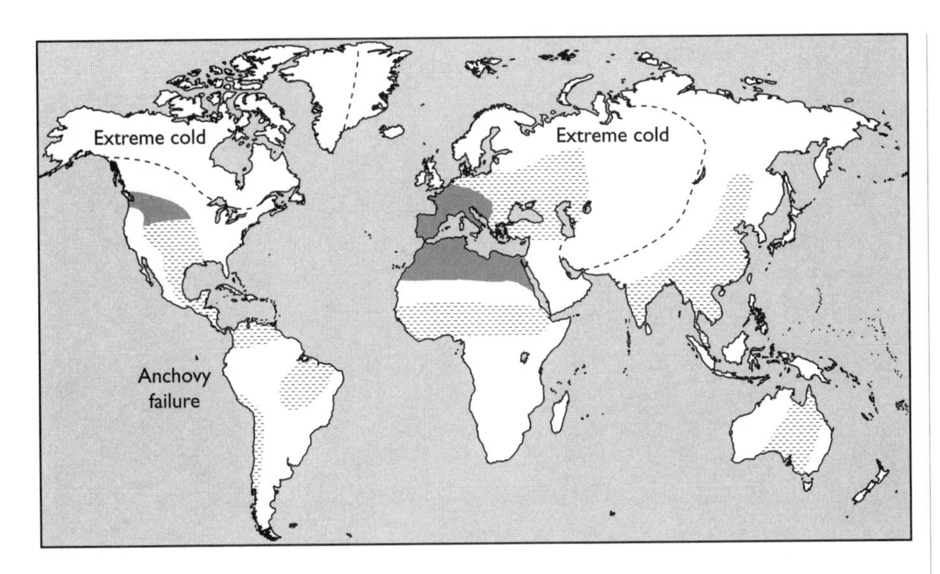

Figure 46 Strange weather of 1972 caused by El Niño. Hatched areas were affected by extreme drought. Areas in dark gray were affected by unusually wet weather.

regions would significantly affect agriculture, forcing it to migrate to higher latitudes. Canada and Russia might then become breadbaskets. Unfortunately, the soils in the northern regions are thin because of glacial erosion during the ice ages and would become quickly depleted with extensive agriculture.

Major groundwater supplies would also be adversely affected, as water tables drop and wells go dry. Higher temperatures would augment evaporation and reduce the flow of some rivers by 50 percent or more, while other streams would dry out entirely. During the 1988 drought, one of the worst of the 20th century, the Mississippi River fell to a record low, making navigation impossible over long stretches. Ancient sunken derelicts became visible for the first time since their sinking.

Other areas could receive a marked increase in precipitation, causing extensive flooding. Relatively modest shifts in the globe's average annual temperature and rainfall could have a dramatic regional effect on the frequency of catastrophic floods. Heavy precipitation would become more frequent and more intense as Earth's atmosphere continues to warm. Greenhouse warming would energize the atmosphere, making storm systems increasingly more violent. Moreover, changing weather patterns due to instabilities in the atmosphere could create deserts out of once productive farmlands. These changing patterns would also drench other regions and cause severe floods and soil erosion. Larger-than-normal seasonal temperature variations might occur along with a higher atmospheric moisture content, producing storms of unprecedented proportions.

The central regions of the continents, which normally experience occasional droughts, would become permanently dry wastelands. The soils in

Figure 47 *Hurricane Diana viewed from space.* (Photo courtesy NOAA)

almost all of Europe, Asia, and North America would dry out, requiring additional irrigation. Expected rises in temperatures, increased evaporation, and changes in rainfall patterns would severely limit the export of excess food to developing nations during times of famine. Dry winds of tornadic force would create gigantic dust storms and severe erosional problems. Tornadoes, hailstorms, thunderstorms, and lightning storms would increase in frequency, intensity, and duration. Numerous immense hurricanes would charge headlong into heavily populated coastal areas (Fig. 47), resulting in tremendous property damage and great losses of life.

CLIMATIC EFFECTS OF VOLCANOES

Volcanoes have a huge influence on the climate over large parts of the world. Many of the coldest and wettest climates have been termed *volcanic dust years.* Volcanic dust probably played some part in all the worst summers and coldest winters from the 17th through the 20th centuries (Fig. 48). The severe winter of 1783–84 was caused by volcanic eruptions in Iceland and Japan. The eruptions of Tambora in Indonesia in 1816 produced a "year without summer." With more than 200 years of observations of volcanoes and weather since 1784, the correlation is quite close, although volcanoes alone cannot explain all the disturbances of the climate.

The British meteorologist Hubert H. Lamb surveyed all volcanic eruptions from the years 1500 to 1970. He related their impact on the atmosphere to a standard scale using the 1883 eruption of Krakatoa as 1,000 units on the dust veil index. The 1816 eruption of Tambora blasted three times more dust into the upper atmosphere as did Krakatoa; therefore, its dust veil index would be 3,000. Since Tambora was proceeded and followed by other eruptions around the world between 1811 and 1818, together their total veil was 4,200—the largest in modern history.

The British meteorologist P. M. Kelly (a former student of Lamb), and his colleagues at the University of East Anglia in England looked for regular variations in the dust veil index and climate. They found that both indicated a 7- to 8-year cycle from 1725 to 1950. This discovery supported the idea that changing volcanic activity influences the climate. The dust veil alone could not be responsible for all the climatic changes over the past centuries, however.

Although volcanic dust is an important contributor to climatic change, it might not be the only cause nor even the main cause of some climatic variations

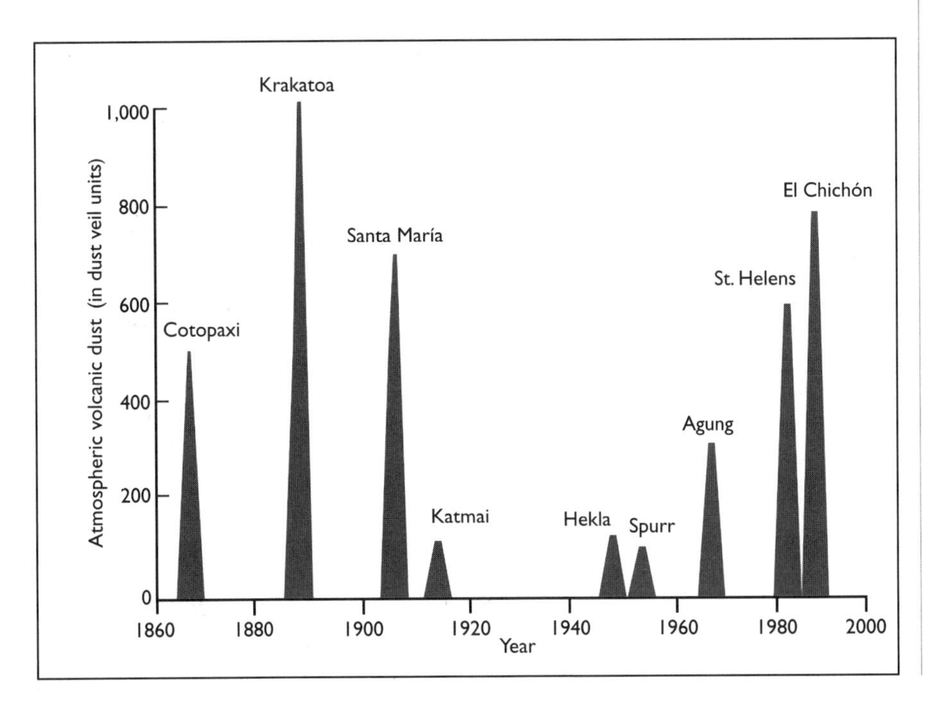

Figure 48 *The relative load of volcanic dust injected into the atmosphere from a sampling of major volcanoes.*

over the centuries. A combination of changing volcanic influence and the changing solar influence measured by sunspot activity might best explain the pattern of climatic change over the past 500 years. Volcanic dust and aerosols still account for the largest influence on the climate. They coincide with the record of temperature changes since 1600. In contrast, sunspot variations do not provide such a good agreement.

Only part of the lost incoming solar radiation is reflected back into space by the volcanic dust (Fig. 49). Some radiation warms the dust itself, while some scatters sideways and reaches the ground at an indirect angle. This scattering of sunlight is responsible for making the sky blue and for causing spectacular sunrises and sunsets. Therefore, scattered solar radiation does not reach the surface directly from the Sun's disk, which is what measurements of direct solar radiation record. If the total output of the Sun did fluctuate by as much as 10 percent, surface temperatures would reflect a much larger fluctuation. A decrease in direct solar radiation of 5 percent would actually cause the surface to cool by less than 1 degree because the decrease in direct radiation is matched by an increase in indirect radiation, scattered sideways by the volcanic dust.

The effect of dust in the atmosphere on the climate depends on the nature of the dust and its location in the atmosphere. Krakatoa was a less significant eruption than Tambora in terms of the effects on climate even though

Figure 49 *The eruption cloud (visible in center of image) from the 1982 eruption of El Chichón, Chiapas, Mexico.*

(Photo courtesy NOAA)

the volcano sent an impressive amount of dust into the atmosphere. Eruptions that throw dust into the atmosphere are divided into two categories. The first includes those that create dust layers in the troposphere and lower stratosphere up to about 20 miles altitude. The second includes those that reach altitudes of about 30 miles or more. The first group has the greatest influence on the climate because such eruptions produce dense, long-lived dust clouds.

The dust alone does not block out heat from the Sun. Volcanoes also produce vast quantities of water vapor and gases. The gas, sulfur dioxide, reacts with water to produce sulfuric acid. These aerosols also penetrate the stratosphere like a fine mist and obscure sunlight. With the combination of both dust and aerosols being injected into the atmosphere, the large volcanic eruptions make the most important contributions affecting the climate.

The American meteorologist Harry Wexler made the connection between volcanoes and climate in the early 1950s. He noted that following a major eruption, a marked change occurred in the weather patterns over North America. July's weather map resembled that of mid-May. Wexler also made the striking discovery that for 50 years since 1912, no major volcanic eruptions have occurred in the Northern Hemisphere. During this period, the winters have steadily warmed, making 20th-century climate conditions comparatively warmer than those of the previous century. The most conspicuous change in the weather patterns is that for 150 years prior to 1912, volcanoes have erupted in the Northern Hemisphere in one great explosion after another. Since 1912, however, they have been relatively quiet.

The climatic effects from volcanic eruptions in the higher latitudes, such as those of Mount St. Helens, Washington, in 1980 (Fig. 50) and Augustine, Alaska, in 1986, are not nearly as significant as those in the lower latitudes. This is because volcanic dust blasted into the stratosphere in the temperate zones tends to spread less and thereby has a smaller effect than dust injected by volcanoes in the Tropics. The dust from volcanic eruptions in the Tropics is carried poleward by a high-altitude flow of air originating from the Tropics and concentrated in the higher latitudes, where sunlight strikes Earth from a steep angle and therefore has a longer path through the dust.

Among the best-documented evidence of how dust from volcanoes affects the transmission of solar radiation through the atmosphere was the 1963 eruption of Mount Agung in Bali, a geographic position much like that of Tambora. After the eruption, the temperature of the troposphere between 30 degrees south and 30 degrees north latitudes fell significantly by late 1964. By the end of 1966, the temperature had recovered to about its preeruption average. The effect was more intense in the northern latitudes, where it amounted to a decrease of about 0.5 degrees Celsius.

The 20th century's answer to Krakatoa as far as the effects on the climate are concerned was the 1982 explosive eruption of El Chichón in southern

Figure 50 *A large eruption cloud from the July 22, 1980, eruption of Mount St. Helens, Skamania County, Washington.*

(Photo courtesy USGS)

Mexico (Fig. 51). The ash cloud might have cooled the Northern Hemisphere by as much as any volcanic eruption during the 100 years following the Krakatoa eruption. A dense cloud of sulfurous gases and dust shot into the stratosphere. The highest concentration was measured above Hawaii at approximately 16 miles altitude. The immense ash cloud was still detectable at least one month

Figure 51 *The decapitated dome of El Chichón from the 1982 eruption, Chiapas, Mexico.*

(Photo courtesy USGS)

Figure 52 *The* Earth Radiation Budget Experiment *(ERBE) satellite measures Earth's albedo for anticipating climate trends that affect agriculture, energy, natural resources, transportation, and construction.*

(Photo courtesy NASA)

after the eruption. It took 10 days to reach Asia, took two weeks to reach Africa, and completely circumnavigated the globe in three weeks. A year later, the average temperature in the Northern Hemisphere dropped about 0.5 degrees Celsius. This resulted in an unusually cool, wet summer and one of the coldest winters on record.

Massive volcanism, such as the June 1991 eruption of the Philippine volcano Pinatubo, possibly the largest blast of the 20th century, might slow greenhouse warming by reflecting sunlight back into space. Instruments aboard the *Earth Radiation Budget Experiment* satellite (Fig. 52) detected a nearly 4 percent increase in sunlight reflected by the atmosphere months after the eruption, which reduced surface temperatures by at least 0.5 degrees Celsius. Although volcanic ash falls out of the atmosphere rather quickly, in a matter of weeks or months, volcanic aerosols, chiefly sulfur dioxide, linger for up to three years or more. During the past 100 years, the global temperature has dropped several tenths of a degree within one to two years following large volcanic eruptions.

VIOLENT STORMS

The temperature difference between the equator and the poles, the engine that powers Earth's weather, has decreased since the late 1800s. This has possibly altered the frequency and intensity of storms. An increase in temperatures induced by higher levels of atmospheric greenhouse gases could contribute substantially toward energizing the atmosphere and add extra power to storm systems. Practically every day, a major storm takes lives and destroys property in some part of the world. When statistics show that the death toll and property damage are on the rise, not only is nature becoming more violent but the world is also becoming more crowded, placing more people in harm's way.

The number of weather-related natural disasters has doubled since the middle of the 20th century. In the United States, 38 severe weather events occurred between 1988 and 1999, with seven severe events occurring in 1998 alone, the most for any year on record. That was also the warmest year since global records began in 1860. Furthermore, the strongest El Niño on record occurred from 1997 to 1998, bringing Pacific seawater temperatures as high as 5 degrees Celsius above normal.

Hurricanes are nature's most spectacular and destructive storms. They occur more commonly in summer and autumn, when the Sun heats the ocean well to the north or south of the equator, where Coriolis forces add spin to air masses. When a hurricane makes landfall, it is accompanied by a tremendous storm surge that wrecks property and erodes beachfronts. Winds of 100 miles per hour or more pile up water on the shore, while low pressure in the eye of the hurricane sucks the water up into a mound several feet high.

Tornadoes strike sporadically and violently (Fig. 53). They generate the strongest of all surface winds and cause more deaths annually in the United States than any other natural phenomenon. They develop in the spring and to a lesser extent in the autumn, when conditions are ripe for the formation of tornadic thunderstorms. However, with a warmer global climate, the number of dangerous tornadoes is expected to rise. With about 700 tornadoes yearly, the United States, particularly the central and southeastern portions of the country—known as tornado alley (Fig. 54), has the world's highest incidence of tornadoes. Australia, with its great central desert, where powerful convective thunderstorms occur, is a close second.

At any given moment, nearly 2,000 thunderstorms are estimated to be in progress over the entire world. Their sheer numbers make them the primary balancers of Earth's heat budget. They occur most commonly in spring and summer, and they infrequently occur in winter. On average, more than 100 Americans are killed and about 250 are seriously injured by thunderstorms every year. Property losses are estimated in the hundreds of millions of dollars annually. Thunderstorms, generated by temperature imbalances in the atmosphere, are violent examples of the upward transfer of heat flow. An extreme example of precipitation and cloud turbulence is the formation of hailstorms (Fig. 55).

Dust storms are expected to escalate as the ground becomes hotter and drier and as winds become blusterous. They present a solid wall of dust that

Figure 53 *A tornado near Denver, Colorado, airport.*

(Photo courtesy NOAA)

blows at speeds upward of 60 miles per hour, reaching several thousand feet altitude and stretching for hundreds of miles. The land is scoured by the winds. Several inches of soil can be airlifted to other regions, even across the ocean. Giant dust storms arise in the deserts of Africa, Arabia, central Asia, Australia, and the Americas. The most obvious threat of dust storms is soil erosion. Because of this, each year the deserts claim more valuable land.

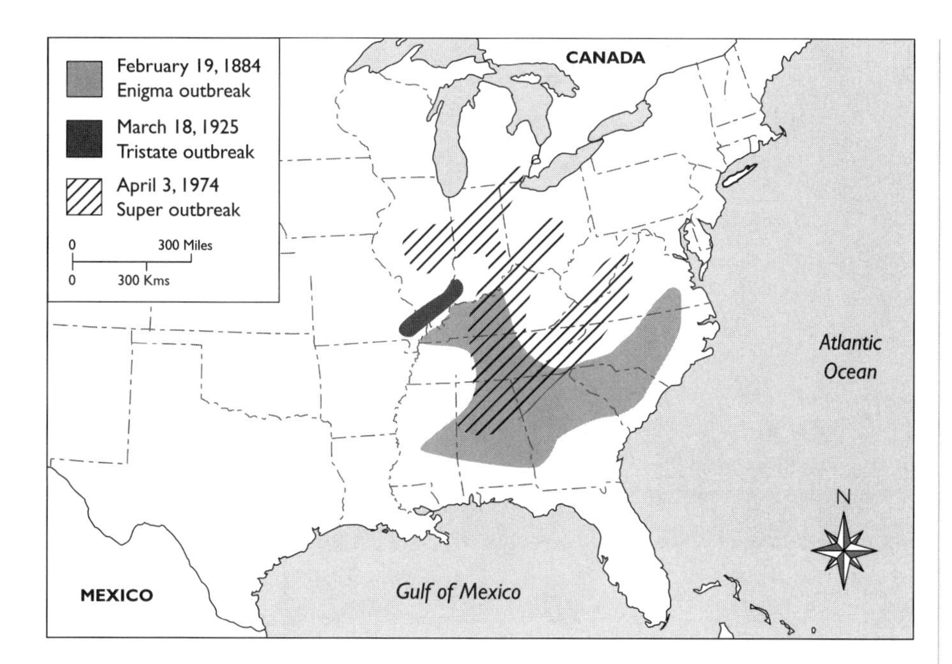

Figure 54 Approximate areas covered by three of the greatest tornado outbreaks. The northeast-trending zones also represent the paths of the tornadoes, which are influenced by the jet stream.

Of all of nature's violence, nothing can compare to lightning for its instantaneous release of intense energy (Fig. 56). Lightning is very destructive to structures and causes most forest fires. No other weather phenomenon kills

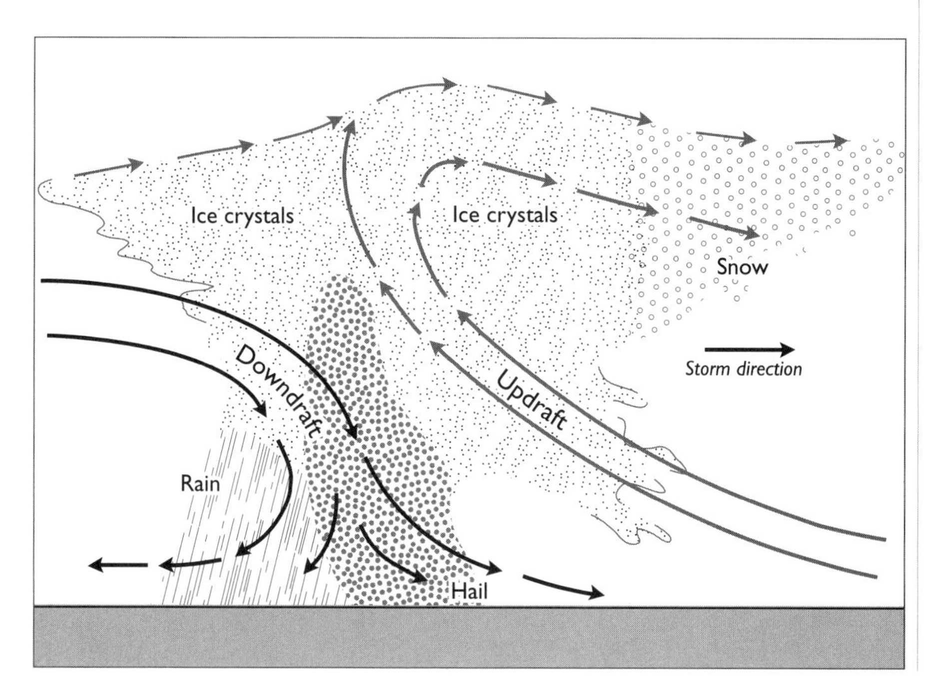

Figure 55 The structure of a hailstorm.

Figure 56 *Tremendous lightning bolts are the major cause of forest fires in the United States.*

(Photo courtesy NOAA)

as many people. Over the past decades, an average of 100 people a year have been killed by lightning in the United States. As the atmosphere becomes more turbulent due to greenhouse warming, the number of lightning strikes is expected to increase. This adds an additional danger of forest fires set during dry conditions.

COMBATING CLIMATE CHANGE

Thus far, the climate changes thought to be related to human activities have been relatively modest. However, the degree of change could become dramatic by the middle of this century, exceeding anything seen in nature during the past 10,000 years. Although some regions might temporarily benefit from higher global temperatures, overall the alterations are expected to be severely disruptive.

Conservation is the preferred method for combating climate change. It would result, in large part, from improved energy efficiency along with the development of nonpolluting substitute energy sources. However, more conservation can attack only part of the carbon dioxide problem, not solve it. Carbon dioxide might also be disposed of in special storage sites such as brine-filled aquifers, deep coal seams, depleted oil formations, and the deep-ocean bottom. On the ocean floor more than 2 miles deep, liquid carbon dioxide does not mix with seawater but forms marble-sized blobs with a consistency of Jell-O.

The theory of climate change has so many complex variables that predicting the weather well into the future is extremely difficult. Since the climate has always been changing, even from year to year often bringing weather extremes (Table 4), any future changes could still result from natural causes. Thus far, no clear sign of greenhouse warming has occurred. Because the present effects of global climate change seem relatively mild, no major actions or sacrifices are deemed necessary at present.

More research is needed on atmospheric physics and air-sea interactions, using the most powerful computers to model the data. Much information about Earth is collected by advanced space technology. Because the amount of information amassed is so huge, a decade or more might be required to analyze the data. Perhaps, if an upward trend in increasing temperature continues well into this century, the climate change can definitely be blamed on the greenhouse effect.

To determine whether greenhouse gases are actually warming the planet, scientists are studying the speed at which sound waves travel through the ocean. Since sound travels faster in warm water than in cold water—a phenomenon known as acoustic thermometry—long-term measurements could reveal whether global warming is a certainty. The idea is to send out low-frequency sound waves from a single station and monitor them from several listening posts scattered around the world. The signals take several hours to reach the most distant stations. Therefore, shaving a few seconds off the travel time over an extended period of five to 10 years could definitely indicate that the oceans are indeed warming.

TABLE 4 THE WARMEST, WETTEST, AND WINDIEST CITIES IN THE UNITED STATES

Extreme	Location	Average annual value
Warmest	Key West, Florida	Mean temperature 78°F
Coldest	International Falls, Minnesota	Mean temperature 36°F
Sunniest	Yuma, Arizona	348 sunny days
Driest	Yuma, Arizona	2.7 inches of rainfall
Wettest	Quillayute, Washington	105 inches of rainfall
Rainiest	Quillayute, Washington	212 rainy days
Cloudiest	Quillayute, Washington	242 cloudy days
Snowiest	Blue Canyon, California	243 inches of snowfall
Windiest	Blue Hill, Massachusetts	Mean wind speed 15 mph

The first direct measurements of a possible ocean warming from rising global temperatures were taken by satellites. The polar sea ice appears to have shrunk by as much as 6 percent since the 1970s when measurements were first made. However, an extensive study of temperatures over the Arctic Ocean indicates the region has not warmed significantly over the last four decades. Perhaps the Arctic is the last region affected by greenhouse warming. Yet climate models suggest that the Arctic would be a good place to predict climate changes because global warming caused by rising carbon dioxide levels and other greenhouse gases should be amplified there. The melting of the Arctic ice pack, like ice cubes dissolving in a cold drink, would not significantly raise the level of the ocean.

Unknown moderating factors might cancel or at least lessen the greenhouse effect. Gaseous sulfur produced by marine single-celled plants called plankton might help counter human-induced global warming by partially regulating Earth's temperature. The sulfur gas emissions could increase the concentration of cloud-forming particles. These could make clouds whiter and therefore more reflective, which in turn would lower global temperatures. Volcanic eruptions, decreasing solar activity, and decreasing stratospheric concentrations of ozone could induce some additional cooling.

Figure 57 *Attempts to rebuild beaches at Ocean City, Worcester County, Maryland.*

(Photo by R. Dolan, courtesy USGS)

Large-scale human intervention might be required to preserve plant and animal species threatened by global climate change, especially if the change happens too quickly. The two response strategies for combating climate change are adaptation and limitation. Adaptation might involve anything from moving to a cooler climate to building coastal defenses against a rising sea (Fig. 57). Limitation directly involves limiting or reducing the emissions of greenhouse gases. Perhaps a prudent response to climate change would be the implementation of both these measures.

If, however, corrective measures are not forthcoming, much more drastic steps might be required to counter future global warming. Also, lead times for building greenhouse-combating projects, such as nuclear or solar electrical power plants, might require a decade or more. The thermal inertia of the ocean could delay the onset of greenhouse warming by several decades, by which time the effects of global warming might be catastrophic.

After discussing how greenhouse gases and other pollutants in the atmosphere can change the climate, the next chapter shows how the climate works toward distributing water over the land and the effects and control of flooding.

4

HYDROLOGIC ACTIVITY

WATER FLOW AND FLOODING

This chapter examines the importance of hydrology and the effects of flooding on people and their property. No other substance is more important to life on Earth as water. This is the only planet known to contain water in solid, liquid, and gaseous states. Freshwater represents only about 2.5 percent of all water on the planet, sufficient to fill the 1-mile-deep Mediterranean basin 10 times over. Three-quarters of the freshwater is contained in glacial ice at the poles and atop many mountains.

The remaining less than 1 percent of Earth's water is atmospheric water vapor, running water in rivers, standing water in lakes, groundwater, soil moisture, and water contained in plant and animal tissues. Nearly all Earth's liquid freshwater is hidden out of sight as groundwater, which dwarfs that found in lakes, rivers, and streams. Of all the world's freshwater not locked up in glacial ice, 97 percent consists of groundwater.

Water is one of the most valuable natural resources. People heavily rely on it for industry, agriculture, and urbanization. Unfortunately, it is too often taken for granted and is wasted or polluted. Even advanced technology to find, transport, and conserve freshwater might not be able to accommodate rising demands brought on by soaring human populations. A serious water crisis might result in armed conflicts, retard economic progress, and devastate populations.

THE HYDROLOGIC CYCLE

Water is a unique substance. A water molecule is composed of an oxygen atom and two hydrogen atoms. The hydrogen atoms are separated by electrical charges at an angle of about 105 degrees. The oxygen atom has a weak negative charge, and the hydrogen atoms each have a slight positive charge. These charges allow water molecules to clump together, forming groups of up to eight molecules. More groups form near the freezing point. Because these groups take up more space than individual molecules, water expands when frozen, just the opposite of what most natural materials do. Ice is therefore less dense and able to float on water. This is a fortunate characteristic because if ice were allowed to sink to the bottom of the ocean, which is already near freezing, it could accumulate and fill the entire ocean basin to become a solid block of ice.

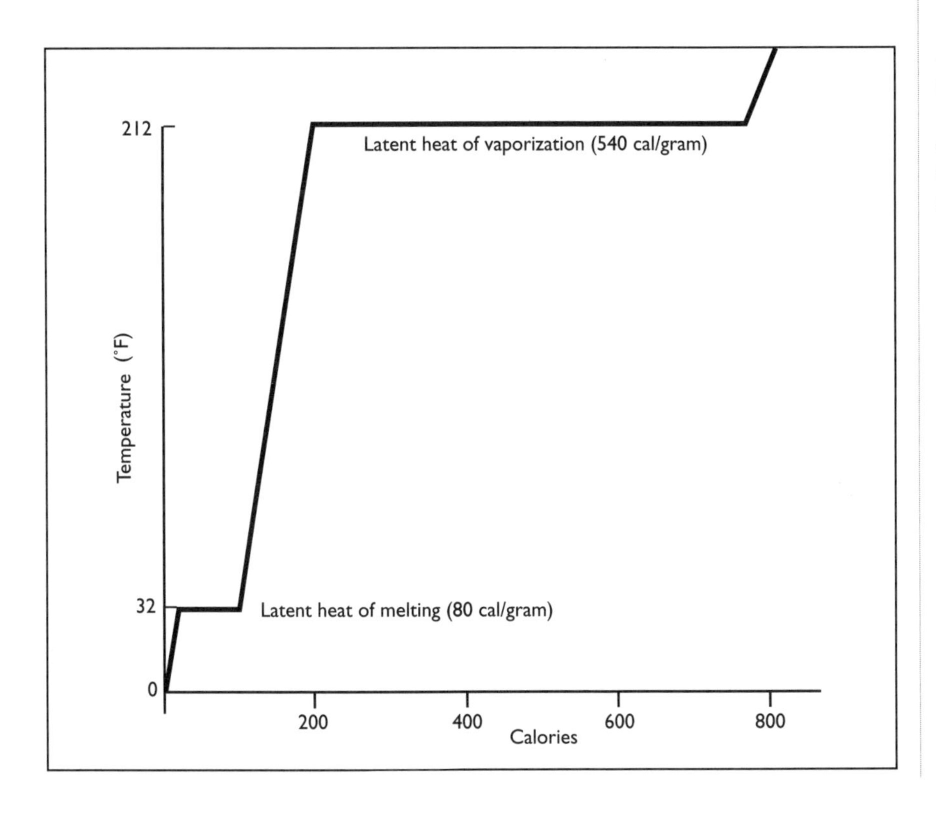

Figure 58 *The change in state of water. To melt 1 gram of ice requires 80 calories. To vaporize that water requires 540 calories.*

Water has the highest specific heat, or heat capacity, of any natural substance. Eighty calories of heat are required to melt 1 gram (0.035 ounces) of ice, known as the latent heat of melting. Another 100 calories of heat are needed to raise the temperature to the boiling point, called sensible heat. Additionally, 540 calories of heat are used to turn water into vapor, known as the latent heat of vaporization (Fig. 58). Therefore, the conversion of water vapor back to 1 gram of ice liberates a total of 720 calories of heat.

Clouds form when a moisture-laden parcel of air is heated and rises through the atmosphere. As the air slowly ascends, the atmospheric pressure gradually decreases, making the air expand. The energy needed for this expansion originates from within the parcel of air itself as heat, resulting in a drop in temperature. The loss of temperature with altitude is called the adiabatic lapse rate (Fig. 59), which is about 1 degree Celsius per 100 meters or about 5 degrees Fahrenheit per 1,000 feet.

If the parcel of moist air continues to rise, it eventually reaches the dew point at which time water vapor then condenses. As the water vapor condenses, it releases latent heat, which is given off when water changes state from a gas to a liquid or from a liquid to a solid. The release of latent heat slows the cooling of the air. The air thus maintains its buoyancy, which contributes to

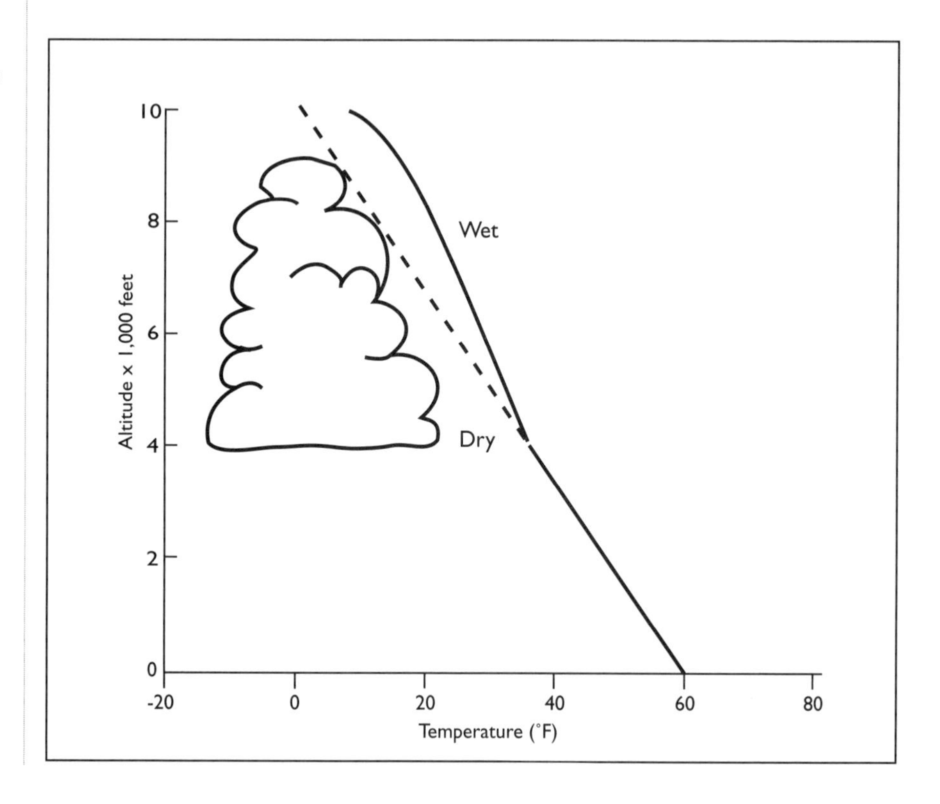

Figure 59 *The adiabatic lapse rate, which is 1 degree Celsius per 100 meters or 5 degrees Fahrenheit per 1,000 feet.*

the upward growth of the cloud because moist air is more buoyant than dry air. Once the saturation point is reached, the air increases buoyancy. This explains why clouds accompany ascending air masses in low-pressure systems.

The atmosphere holds about 0.5 percent of Earth's water at any given moment. Most of the moisture in the air evaporates from the ocean, and about 15 percent originates from the land. Some 15,000 cubic miles of water evaporate from lakes, rivers, aquifers, soils, and vegetation annually. More water evaporates at higher global temperatures. This increases atmospheric water vapor, which happens to be the most effective greenhouse gas. However, the expanded cloud cover resulting from the additional water vapor would probably balance out much of the greenhouse warming.

The greenhouse effect tends to warm Earth. This makes the atmosphere more turbulent and invigorates the hydrologic cycle, producing more intense storm systems. Thunderstorms, generated by temperature imbalances in the atmosphere, are violent examples of the upward transfer of heat flow, which can increase during global warming. The vast numbers of these storms make them the primary balancers of Earth's energy budget. A higher frequency of storms from greenhouse warming would also increase flood hazards in certain regions of the world.

The movement of water over Earth is one of nature's most important processes, known as the hydrologic cycle (Fig. 60). Without it, life as it is now could not exist. The average journey of water from the ocean to the atmosphere, across the land, and back to the sea takes about 10 days. The journey is only a few hours long in tropical coastal areas but can take up to 10,000 years in the polar regions. The quickest route by which water travels to the ocean is by stream runoff. A slower means by which water returns to the sea is by groundwater flow. The slowest route is by glacial flow, as snow accumulating in the polar regions builds glaciers that plunge into the sea.

The oceans cover about 70 percent of the planet's surface to an average depth of more than 2 miles. The total amount of seawater is about 300 billion cubic miles. Daily, 1 trillion tons of water rains down onto the planet, most of which falls directly back into the sea. The average annual rainfall over the entire Earth's surface is roughly 25 inches. Some areas receive more rainfall than others, which determines their geography. The desert regions generally receive less than 10 inches of rainfall yearly, whereas the annual rainfall in the tropical rain forests is upward of 400 inches. The total precipitation on land is about 25,000 cubic miles of water annually. Of this, some 10,000 cubic miles is surplus water lost by floods, held by soils, or contained by lakes and wetlands.

About a third of the surface water is base flow, which is the stable runoff of all the world's rivers and streams. The remainder is groundwater flow, of which only about 1 percent manages to reach the sea. Groundwater leaks into the ocean as coastal springs bubbling directly into the sea or by a process called

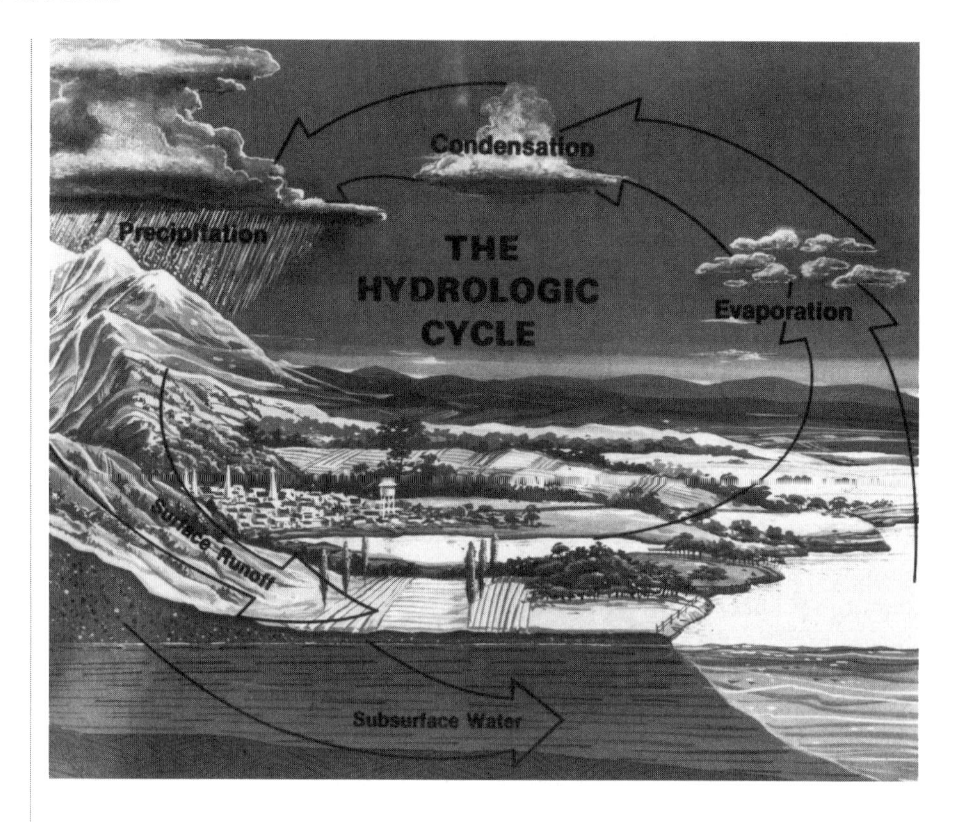

Figure 60 *The hydrologic cycle is a continuous flow of water moving from the ocean over the land and returning to the sea.*

(Photo courtesy USGS)

tidal pumping, which extracts water from aquifers during an ebbing tide. The latter method is how most groundwater enters the ocean. All water falling onto the land surface eventually flows back to the sea, completing the final and most important leg of the hydrologic cycle.

HYDROLOGIC MAPPING

Hydrologic mapping provides critical data on river flow, flood inundation, snow cover, and sea ice extent. The data are used for monitoring flash floods from large storm systems. Satellite-derived precipitation estimates and trends aid meteorologists and hydrologists in evaluating heavy precipitation events and providing timely warnings to affected areas.

Despite the construction of flood-prevention projects to help save lives and reduce property losses, flood damage in the United States often exceeds $1 billion annually. To reduce flood-related hazards, engineers and governmental officials need accurate information on the location of flood-hazard areas and assessments of areas of inundation when floods occur. Computer models are used to provide quick approximations of the total extent of a flood for disaster and relief planning.

Regional snow cover maps are important for predicting runoff during spring thaws. Weather satellite images are used to monitor river basin snow cover in the United States and Canada. Several governmental agencies and private concerns such as utility companies use the river basin snow maps to determine water availability. The snow data also aid in dam and reservoir operations as well as help to calibrate runoff models. These models are designed to simulate and forecast daily stream flow in basins where snowmelt is a major contributor to runoff. This is particularly important in the American West for preparing seasonal water supply forecasts.

The snow cover maps display the areal extent of continental snow fields but do not indicate the snow depth, which is obtained manually in the field. The snow cover charts are digitized and stored in computer databases, from which are created monthly anomaly, frequency, and climatological snow cover maps. In addition, continental or regional snow cover can be calculated over long periods for North American winter snow cover.

The data are useful for detecting and locating ice cover and ice dams on rivers, especially northern rivers, where ice is particularly troublesome. Often, the ice persists because of dams, sharp bends in the river course, or branching of the main channel by islands. Observation of river ice is important because it creates problems for hydroelectric dams, bridges, and maritime navigation. The ice becomes particularly hazardous when it breaks up and forms ice dams, posing a flood threat to nearby communities (Fig. 61).

RIVER FLOW

Rivers are continuously evolving and adapting to environmental pressures, including human modifications. Rivers and streams provide waterways for commerce and water for irrigation, hydroelectric power, and recreation. Running water is responsible for changing the landscape more than any other natural process. Rivers carve out a rugged landscape and are the primary agents for transporting the products of erosion. Surface runoff cleanses the land and supplies minerals and nutrients to the sea. Only a small fraction of the freshwater supply is available for human needs, however, most of which is used for agriculture.

A drainage basin comprises the entire area from which a stream and its tributaries receive water. For example, the Mississippi River and its tributaries drain an enormous section of the central United States from the Rockies to the Appalachians (Fig. 62). Furthermore, all tributaries emptying into the Mississippi have their own drainage areas, becoming parts of a larger basin.

Figure 61 *An ice jam in the Passumpsic River, causing flooding at St. Johnsbury Center, Vermont.*

(Photo courtesy USDA Soil Conservation Service)

Rivers transport sediments delivered to the stream channel by tributaries and by slope erosion on the valley sides. The sediments carried by rivers are temporarily stored by deposition in the channel and on the adjacent floodplain. Streams, heavily laden with sediments, overflow their beds, forcing them to detour as they meander toward the sea.

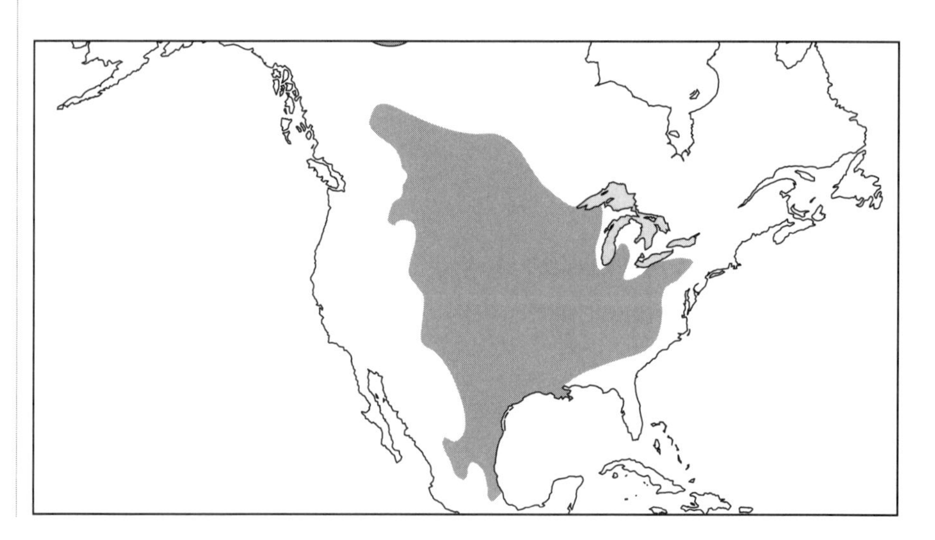

Figure 62 *Major river basins in North America.*

The river content consists of suspended load, bed load, and dissolved load. The suspended load is fine material that settles out very slowly and is therefore carried long distances. The total amount of sediment in suspension increases downstream as more tributaries enter the river. The suspended load is nearly two-thirds the total river content and amounts to about 25 billion tons per year. When rivers reach the ocean, their velocity falls off sharply. Their sediment load drops out of suspension as the sediments continually build the continental margins outward.

The bed load contains material such as pebbles and boulders that travel by rolling and sliding along the river bottom during high flow or flood and is one-quarter or less of the total river content. The dissolved load derived from chemical weathering and from solution by the river itself comprises about 10 percent of the total river content. Most of the dissolved matter in a stream originates from groundwater draining from a breached water table. Groundwater holds a higher concentration of dissolved materials than rivers and streams. Minerals such as limestone dissolve in slightly acidic river water. Limestone also acts as a buffer to maintain acidity levels within tolerable limits for aquatic life.

River deposits called alluvium accumulate because of a decline in stream gradient or inclination, a reduction in stream flow, or a decrease in stream volume, with the heaviest material settling out first. Changes in the river environment occur when entering standing water, encountering obstacles, evaporation, and freezing. River deposition is divided into deposits in bodies of water, alluvial fans, and deposits within the stream valley itself. A medium-sized river will take about a million years to move its sandy deposits 100 miles downstream. Along the way, the grains of sand are polished to a high gloss. When reaching the ocean, the grains become beach sands reworked by wave action.

Sedimentary rocks deposited within streambeds are relatively rare because rivers deliver most of their sediment load to lakes or the sea. River deltas (Fig. 63) develop where streams enter larger rivers or standing water. The velocity of the river slows abruptly when entering a body of water, causing its bed load to deposit onto the bottom. Much of the river's load is also reworked by offshore currents, creating marine or lake deposits.

Channel fill is alluvium laid down in the channel of a stream. Accumulations of fill take many shapes generally known as sandbars. They collect along the edges of a stream, especially on the insides of bends, accumulate around obstructions, and pile up into submerged shoals and low islands. These deposits are not permanent features but are either destroyed, redeposited, or shift positions as conditions on the river change.

Alluvial fans generally found in arid regions are similar to river deltas. They form where streams flow out of mountains onto broad valleys, where

Figure 63 *Delta of the Chelan River entering the Columbia River, Chelan Ferry, Washington.*

(Photo by B. Willis, courtesy USGS)

the ground abruptly flattens. This causes the stream to slow and deposit its sediment load in a fan-shaped body. As the stream constantly shifts its position, the alluvial fan grows steeper, thicker, and coarser, developing a characteristic cone shape.

As a river excavates its floodplain due to changes in flow, it leaves a terrace standing above the river's new level. Terraces first develop in the lower reaches of a stream and extend upstream, cutting into sediments laid down earlier in either direction. Terraces also form by lateral cutting of bedrock by a river. When rivers produce more debris than they can handle, the excess sediment is deposited into the river valleys, which form terraces when rivers downcut these deposits.

Another type of river flow called a braided stream (Fig. 64) forms when the bed load is too large and coarse for the slope and the amount of discharge. The banks easily erode. This chokes the channel with sediment, causing the stream to divide and rejoin repeatedly. The stream deposits the coarser part of its abundant load to attain a steep enough slope to transport the remaining load. This forces the stream to broaden and erode its banks. Alluvium rapidly deposits in constantly shifting positions, forcing the stream to split into interlacing channels that continuously separate and reunite.

As rivers clogged with sediment fill their channels and spill over onto the adjacent plain, they carve out a new river course. In the process, they meander downstream, forming thick sediment deposits in broad floodplains that can fill an entire valley. As a stream meanders across a floodplain, the greatest erosion takes place on the outside of the bends, resulting in a steep cut bank in the channel. In contrast, on the inside of the bends, the water slows and deposits its suspended sediments. During a flood, a winding river often takes a shortcut across a low-lying area separating two bends. The river temporarily straightens until it further fills its channel with sediment, causing it to meander again. Meanwhile, the cutoff sections of the original river bends become oxbow lakes.

The largest rivers of North and South America empty into the Atlantic Ocean. Because it is smaller and shallower than the Pacific, its waters tend to be saltier. Along the eastern coast of the Americas, the sea lies on a wide and gently sloping continental shelf that extends eastward more than 60 miles, reaching a maximum depth of 600 feet at the shelf edge. In contrast, along the western coast, the water descends rapidly to great depths a short distance from shore. The sea off the east coast is dominated by freshwater discharges from the great coastal plain estuaries from the St. Lawrence River in the north to the

Figure 64 *Braided channels of the Nelchina River, Copper River region, Alaska.*

(Photo by J. R. Williams, courtesy USGS)

great Amazon River in the Tropics. In marked contrast, the west coast is dominated by periodic upwelling of bottom water along the Oregon coast to the coast of Peru.

Major rivers such as the Amazon and the Mississippi transport enormous quantities of sediment derived from their respective continental interiors. The Amazon, the world's largest river, is forced to carry heavier sediment loads due to large-scale deforestation and severe soil erosion at its headwaters. The Mississippi River dumps hundreds of millions of tons of sediment into the Gulf of Mexico annually, widening the Mississippi Delta and slowly building up Louisiana and nearby states (Fig. 65). The Gulf Coastal states from east Texas to the Florida panhandle were built up with sediments eroded from the interior of the continent and hauled in by the Mississippi and other rivers.

On a global scale, about 40 billion tons of sediment are carried by stream runoff into the ocean annually. India's Ganges River carries four times more sediment than the Amazon River, which is three times larger. The Ganges and Brahmaputra Rivers convey about 40 percent of the world's total sediment

Figure 65 *Sediment deposition in the Mississippi River Delta: 1930 conditions (left), 1956 conditions (right).* (Photo by H. P. Guy, courtesy USGS)

discharge into the ocean, as erosion gradually wears down the Himalaya Mountains, whose remains are dumped into the Bay of Bengal. This creates a gigantic sediment pile as much as 3 miles thick.

GROUNDWATER FLOW

Groundwater is the second most important source of freshwater. A groundwater aquifer consists of unconsolidated sand and gravel sandwiched between impermeable layers. Water flows through this formation by gravity at a very slow rate. Water catchment areas at the head of the aquifer recharge the groundwater system. These areas must remain undeveloped to operate properly. The rate of infiltration into the groundwater system depends on the distribution and amount of precipitation, the type of soils and rocks, the slope of the land, the amount and type of vegetation, and the quantity of water rejected because the ground is oversaturated with soil moisture.

Groundwater aquifers are being pumped down faster than they are naturally replenished in many parts of the world, especially in heavily populated nations such as China, India, and the United States. Many aquifers in the American West contain Pleistocene water. They were recharged during the ice ages over the last 2 million years or so. The overuse of groundwater tends to lower the water table or deplete the aquifer altogether. Once an aquifer is pumped dry, it can no longer be restored to its original capacity if subsidence caused by the weight of the overlying strata compresses the water-bearing sediments. The compaction decreases the pore spaces between grains, through which the water must flow. Often, subsidence results from the depletion of groundwater aquifers, causing the ground to drop several feet in many places.

Water flows from the recharge area through formations of porous sand and gravel at a maximum rate of only a few inches per year. The mistaken belief that groundwaters are bountiful subterranean rivers is dispelled as one aquifer after another is depleted by overpumping. Groundwater that manages to reach the ocean forms a freshwater to seawater interface near the shore. However, excessive groundwater use in coastal areas can cause the loss of hydrostatic head, allowing saltwater intrusion to contaminate wells, rendering them useless.

In the United States, more than half the population depends on groundwater for domestic use and irrigation (Fig. 66). Some 100 billion gallons of water per day are pumped from groundwater aquifers. Many midwestern and western states draw more than half their water from the ground. In several million rural households, aquifers are the only source of water. Most of the water is pumped from wells less than a few hundred feet deep, making them susceptible to pollution. Groundwater contamination in these areas could be

Figure 66 *The discharge of an irrigation well near Victor, California.*

(Photo by H. T. Stearns, courtesy USGS)

catastrophic because of an inadequate source of surface water for domestic, industrial, and agricultural needs.

IRRIGATION

More than 10 percent of the world's cropland is irrigated, requiring about 600 cubic miles of water annually. Groundwater irrigation is also very expensive. Only affluent nations can afford it on a large scale. The United States irrigates nearly one-quarter of its farmland and has tripled the amount of irrigated land over the last half century. Irrigation has many advantages as well as severe drawbacks. Overuse of groundwater depletes aquifers, possibly causing subsidence, and severely limits the recovery of groundwater systems. Most river water used for irrigation has a high salt content. If fields are not drained properly, salt buildup in the soil can ruin the land, stunting or killing crops. Tens of thousands of acres of once fertile land are destroyed by this process annually. If trends continue, possibly within just a few decades, over half of all irrigated land will be rendered useless by soil salinization.

Most of the eastern third of the nation receives sufficient rainfall to support agriculture without irrigation. In contrast, much of the western portion of the country is rain deficient and must be supplemented by irrigation. Heavy use of irrigation, which not long ago turned vast stretches of America's western desert into the world's most productive agricultural land, is now ruining hundreds of thousands of acres.

Irrigation water gradually degrades the land by the buildup of salts such as sodium, calcium, and magnesium chlorides. At least one-third of California's farmland is in danger of being destroyed by salt. Lower water availability in the spring and summer also could dramatically reduce crops. Meanwhile, the land in some regions contains selenium, arsenic, boron, and other naturally occurring poisons that are polluting the runoff.

With most of the good irrigated land already in production, farmers cannot afford to cultivate the land until it salts up and has to be abandoned. The problem is exacerbated in arid regions because of the high natural salt content of the soil and low amounts of rainfall needed to flush out the excess salts. Agricultural chemicals such as fertilizers and pesticides are carried off by the irrigation drain water. This is dumped into rivers, which finally reach the ocean, where high concentrations of chemicals can kill fish and other aquatic life.

Irrigation runoff has polluted marshes, rivers, lakes, and estuaries in California, Colorado, and other western states. Selenium leaching from soil on heavily irrigated farms in the San Joaquin Valley of California has caused deformities in nearby waterfowl, including twisted beaks, stubs for wings, and missing eyes. The contaminated birds lay thin-shelled eggs that can easily break, threatening whole generations of waterfowl.

FLOOD HAZARDS

Floods are naturally recurring events. They are important geological processes that alter the courses of rivers and distribute soils over the land. Heavily sedimented rivers also increase the severity of floods. Whenever a major flood occurs, it often reshapes the landscape through which the river flows. While a flood is in progress, a river might alter its course several times as it rushes to the sea.

More than 3 million miles of rivers and streams flow across the continental United States. Much of the land area next to these river courses is prone to flooding (Fig. 67 and Table 5). Because many cities are located near streams, a high percentage of the nation's population and property concentrates in flood-prone areas. More than 20,000 communities have flood problems. Due to high population growth, modern floods have become increasingly hazardous as people continue to move into the floodplains.

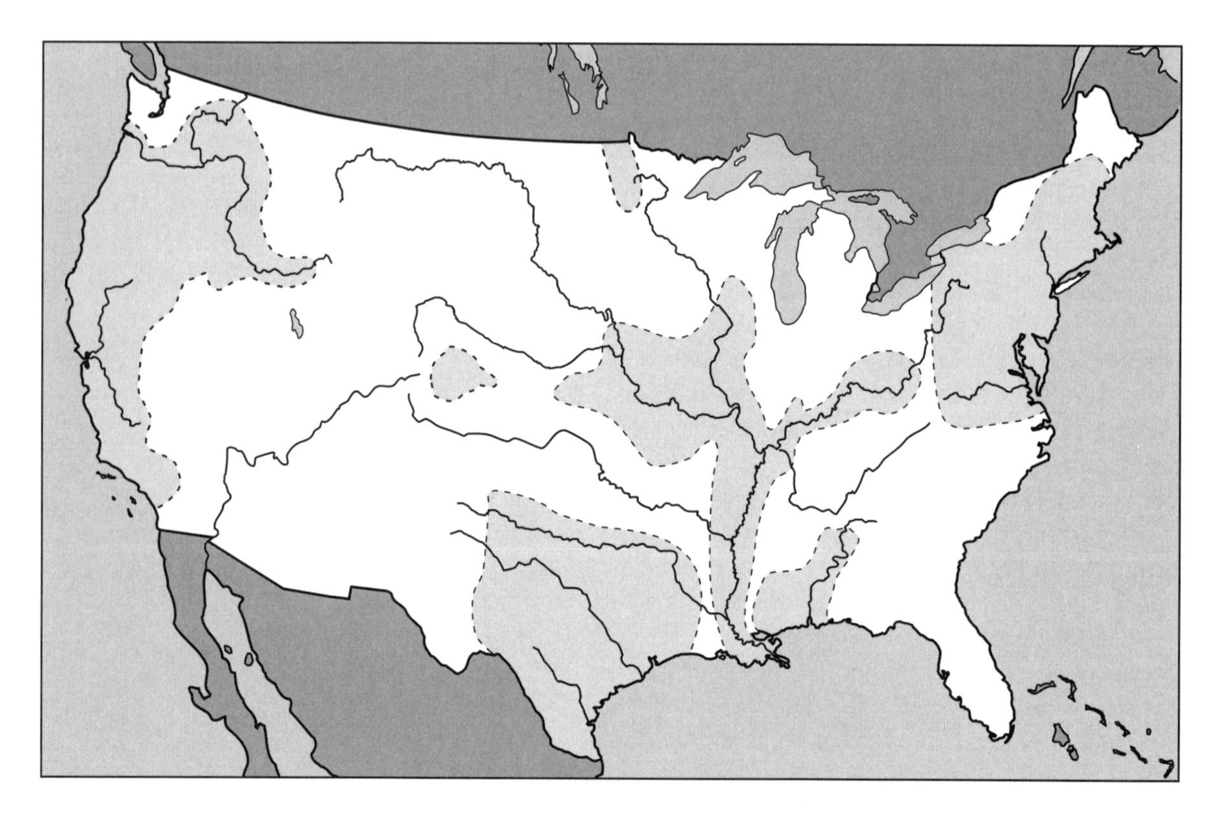

Figure 67 *Flood-hazardous areas in the United States.*

TABLE 5 CHRONOLOGY OF MAJOR U.S. FLOODS

Date	Rivers or basins	Damage (in $millions)	Death toll
1903	Kansas, Missouri, & Mississippi	40	100
1913	Ohio	150	470
1913	Texas	10	180
1921	Arkansas River	25	120
1921	Texas	20	220
1927	Mississippi River	280	300
1935	Republican & Kansas	20	110
1936	Northeast U.S.	270	110
1937	Ohio & Mississippi	420	140
1938	New England	40	600
1943	Ohio, Mississippi, & Arkansas	170	60
1948	Columbia	100	75
1951	Kansas & Missouri	900	60
1952	Red River	200	10
1955	Northeast U.S.	700	200
1955	Pacific Coast	150	60
1957	Central U.S.	100	20
1964	Pacific Coast	400	40
1965	Mississippi, Missouri, & Red Rivers	180	20
1965	South Platte	400	20
1968	New Jersey	160	—
1969	California	400	20
1969	Midwest	150	—
1969	James	120	150
1971	New Jersey & Pennsylvania	140	—
1972	Black Hills, S. Dakota	160	240
1972	Eastern U.S.	4,000	100
1973	Mississippi	1,150	30
1975	Red River	270	—
1975	New York & Pennsylvania	300	10
1976	Big Thompson Canyon	—	140
1977	Kentucky	400	20
1977	Johnstown, Pennsylvania	200	75
1978	Los Angeles	100	20

(continues)

TABLE 5 *(CONTINUED)*

Date	Rivers or basins	Damage (in $millions)	Death toll
1978	Pearl River	1,000	15
1979	Texas	1,250	—
1980	Arizona & California	500	40
1980	Cowlitz, Washington	2,000	—
1982	Southern California	500	—
1982	Utah	300	—
1983	Southeast U.S.	600	20
1993	Midwest U.S.	12,000	24
1997	Red River, North Dakota	1,000	—
1999	Tar River, North Carolina	6,000	—

Lateral migration of bends of rivers and overbank flow combine to produce the floodplain, which is periodically inundated by water and sediment during floods. Channel discharge where water overflows the river bank is called the flood stage. This high-water condition often causes property damage on the floodplain. The purpose of floodplains is to carry off excess water during floods. Only when a flood occurs do people finally recognize the importance of good floodplain management to reduce the severity of floods.

Failure to recognize the function of floodplains has led to haphazard development in these areas with a consequent increase in flood dangers. Floodplains provide level ground, fertile soils, ease of access, and available water supplies. However, because of economic pressures, floodplains are being recklessly developed without full consideration of the flood risk. Consequently, the improper use of floodplains has led to the destruction of property and the loss of life when the inevitable flood arrives.

Floods threaten lives, cause much suffering, damage property, destroy crops, and halt commerce. The average annual flood loss in the United States has increased from less than $100,000 at the beginning of the 20th century to about $4 billion presently. The 1973, 1993, and 2001 Mississippi River floods, the 1978 Pearl River flood in Louisiana and Mississippi, and the 1997 Red River of the North flood were some of the costliest floods in American history. The Red River bordering North Dakota and Minnesota is notorious for flooding during spring melt (Fig. 68). The April 1997 flood drove the river up to 50 feet or more above flood stage, causing more than $1 billion in property damage and forcing some 100,000 people from their homes.

Figure 68 *A flooded farmstead in the Red River Valley, North Dakota, in 1975.*

(Photo by C. Olson, courtesy USDA)

Modern floods have increasingly become man-made disasters because of the continued construction on the floodplains without recognizing the flood potential. Floodplain zoning laws and flood control projects are based on short-term historical records of large floods, often referred to as 50- and 100-year floods. The lack of long-term data from the geologic record generally makes assessing the risk of large floods difficult. Furthermore, due to climate variability, two or more record-breaking floods can occur in consecutive years. Global warming also has the potential of bringing on more record-breaking floods due to increased instabilities in the atmosphere.

Despite flood protection programs, the average annual flood hazard is rising because people are moving into flood-prone areas faster than flood protection projects are being built. The increased losses therefore are not necessarily the result of larger floods but of greater encroachment onto floodplains. As the population increases, more pressure is exerted to develop flood-prone areas without taking proper precautions. People are often uninformed about the flood risk when building in flood-prone areas. When the inevitable flood strikes, they invariably turn to the federal government to pay for rebuilding in the same floodplain.

FLOOD TYPES

Riverine floods are the result of heavy precipitation over large areas, of the melting of winter's accumulation of ice and snow, or a combination of both (Fig. 69).

Figure 69 *Severe flooding at Fairbanks, Alaska, on August 15, 1967.*

(Photo by J. M. Childers, courtesy USGS)

Although the undeveloped countries contain dense populations living on floodplains and lose a higher proportion of lives, they have relatively lower amounts of property damage than developed nations. Even though the number of lives lost to flooding in the United States approaches 100 people annually, yearly property damages average more than $200 million. The low number of deaths compared with poorer countries reflects improved monitoring and warning systems before the flood and disaster relief afterward.

Many unusual weather events are blamed on El Niño, an anomalous warming of the equatorial eastern Pacific Ocean every two to seven years, with a duration of up to two years. Longer and stronger El Niños have occurred during the last two decades than in the previous 120 years. An odd double El Niño occurred back-to-back from 1991 to 1993 and again from 1994 to 1995. A powerful El Niño from 1997 to 1998 produced record-breaking warmth in the Pacific, causing 23,000 deaths and 33 billion dollars in damages around the world. Seawater temperatures were as high as 5 degrees Celsius above normal compared with just a couple degrees for normal El Niños.

The Midwest floods in the United States during the spring and summer of 1993 resulted in large part because of another strong El Niño event. The jet stream remained stationary over the upper Midwest, where it steered strong weather systems into the region. Major rivers including the Mississippi and Missouri overflowed their levees and poured onto adjacent floodplains. The floods left tens of thousands of people homeless and destroyed millions of acres of cropland. The disaster was the costliest in American history, amounting to 15 to 20 billion dollars in damages and the deaths of 48 people.

The flooding was considered a man-made disaster because levees constructed to protect property severely restrict river flow during flood stage. Furthermore, reservoirs built to contain normal floods tend to overflow during massive flooding. Upstream floodplains and wetlands (Fig. 70) act as sponges to soak up excess floodwaters. Levees restrict this function, causing serious flooding downstream.

Riverine floods occur in river systems whose tributaries drain large geographic areas and encompass many independent river basins. Floods on large river systems continue for a few hours to several days. The floods are influenced by variations in the intensity and the amount and distribution of precipitation. Other factors that directly affect flood runoff are the condition of

Figure 70 *Wetlands such as this one in Dodge County, Wisconsin, are essential for waterfowl and other animals.*

(Photo by Ron Nichols, courtesy USDA Soil Conservation Service)

the ground, the amount of soil moisture, the vegetative cover, and the level of urbanization, where impervious pavement prevents the infiltration of water into the ground.

Upstream floods occur in the upper parts of a drainage system. They are produced by intense rainfall of short duration over a relatively small area. They generally do not cause floods in the larger rivers they join downstream. Conversely, downstream floods cover a wide area. They are usually caused by storms of long duration that saturate the soil and produce increased runoff. The contribution of additional runoff from many tributary basins can cause a large flood downstream, which is recognized by the migration of an ever-increasing flood wave with a large rise and fall in discharge.

Floodwater movement downstream is controlled by the river size and the timing of flood waves from tributaries emptying into the main channel. When a flood moves down a river system, temporary storage in the channel reduces the flood peak. As tributaries enter the main channel, the river enlarges downstream. Because tributaries are of different sizes and irregularly spaced, flood peaks reach the main channel at different times, thereby smoothing out the flow as the flood wave travels downstream and eventually discharges into the sea.

Flash floods are among most severe forms of flooding. They are local floods of great volume and short duration. They occur during violent thunderstorms or cloudbursts on a relatively small drainage area over a short period. They are particularly hazardous if the ground is heavily soaked with water from previous rains or when streams are already at capacity. Flash floods also occur following dam breaks or by the sudden breakup of ice jams, causing the rapid release of large volumes of flow.

A major break in the Teton Dam near Newdale, Idaho, on June 5, 1976 (Fig. 71), caused a flood of unprecedented magnitude on the Teton River, lower Henrys Fork, and Snake River. A wall of water up to 16 feet high devastated communities downstream. The rampaging floodwaters carried off large trees and debris from destroyed buildings and other structures. The water spread over an area of more than 180 square miles, damaging about $400 million worth of property.

One of the nation's most disastrous flash floods raged through in the Big Thompson River Canyon east of Rocky Mountain National Park in north-central Colorado on July 31, 1976. Thunderstorms in the canyon area dumped some 10 inches of rain in a 90-minute interval. A deluge of water rushed down the steep slopes overlooking the river and poured into the narrow canyon. The river rapidly rose and flooded the Big Thompson River and its tributaries between Estes Park and Loveland, wiping out several small communities (Fig. 72). The floodwaters carried off buildings, vehicles, and large trees.

Urban areas with well-designed drainage systems that can handle normal high-water levels are totally swamped by a flash flood. The water level rises

Figure 71 *The June 5, 1976, Teton Dam break, which caused extensive flooding downstream.*
(Photo courtesy USGS)

too rapidly for the drains to handle the excess, and they become overloaded with water that overflows into the streets. Also, runoff from intense rainfalls can result in high flood waves that destroy roads, bridges, homes, buildings, and other community developments.

Flash floods from violent thunderstorms produce flooding on widely dispersed streams, resulting in high flood waves. They are particularly common in mountainous areas and desert regions of the American West. Flash floods are a potential source of destruction and a threat to public safety in areas where the terrain is steep, surface runoff rates are high, streams flow in narrow valleys and gullies, and severe thunderstorms are prominent. The discharges quickly reach a maximum and diminish almost as rapidly. The floodwaters frequently contain large quantities of sediments and debris collected as the river sweeps clean the stream channel. They are deposited in streets, cellars, and the first stories of homes and buildings.

Heavy runoff in mountainous regions forms rapidly moving sheets of water that pick up large quantities of loose material, resulting in mudflows that can cause considerable damage. The floodwaters flow into a stream, where the muddy material suddenly concentrates in the stream channel. The dry streambed rapidly transforms into a flash flood that moves swiftly downhill, often with a steep, wall-like front. Mudflows behave as a viscous fluid and often carry a tumbling mass of rocks and large boulders. Heavy rains falling on loose pyroclastic material on the flanks of volcanoes also produce mudflows.

A glacier outburst flood is a sudden release of meltwater from a glacier or subglacial lake and can be quite destructive (Fig. 73). Several underglacier eruptions have occurred in Iceland during the past century. A volcanic eruption under an Icelandic glacier in 1918 unleashed a massive flood of meltwater, or glacial burst called a *jokulhlaup,* a phenomenon known to Icelanders since the 12th century. An underglacier eruption in the sparsely populated southeastern part of the country on September 30, 1996, melted through the 1,700-foot-thick ice cap and sent massive floodwaters and icebergs dashing 20 miles to the sea a month later. In a matter of days, the below-glacier eruption released up to 20 times more water than the flow of the Amazon, the world's largest river. It destroyed telephone lines, bridges, and the only highway running along Iceland's southern coast.

Figure 72 *Destruction of Drake, Colorado, from the July 31, 1976, Big Thompson River flood.*

(Photo courtesy USGS)

Figure 73 *A bridge at Sheep Creek partially buried by debris from a glacier outburst flood, Copper River region, Alaska.*

(Photo by A. Post, courtesy USGS)

FLOOD CONTROL

Flood dangers can be alleviated by taking certain precautions that ultimately save lives and property. The factors that control flood damage include land use on the floodplain, the depth and velocity of the floodwaters, the frequency of flooding, the rate of rise and duration of flooding, the time of year, the degree of ground saturation, the quantity of sediment load being deposited, and the effectiveness of storm forecasting, flood warning, and emergency services.

Direct flood effects include injury and loss of life. They also include damage to buildings and other structures from swift currents, debris, and sediments. In addition, sediment erosion and deposition might cause a considerable loss of soil and vegetation. Indirect flood effects include short-term pollution of rivers, the disruption of food supplies, the spread of disease, and the displacement of people who have lost their homes in the flood.

Flood protection projects include the construction of reservoirs (Fig. 74) with a storage capacity that can absorb increased flow during floods and moderate the flow rates of rivers. The dams also generate hydroelectric power. Their reservoirs provide river navigation, irrigation, municipal water supplies, fisheries, and recreation. However, without proper soil conservation measures in the catchment areas, the accumulation of silt by erosion can severely limit the life expectancy of a reservoir. An unfortunate consequence of dam construction, however, is that freshwater fish species become threatened or endangered, mainly because water withdrawals have destroyed free-flowing river systems where they once thrived. Also, much valuable land, including that used

Figure 74 *Hoover Dam and Lake Mead on the border of Nevada and Arizona.*

(Photo by W. O. Smith, courtesy USGS)

in agriculture, is inundated. Some 3,000 natural and artificial reservoirs inundate about 120 million acres of land and hold more than 1,500 cubic miles of water, or as much as Lake Michigan and Lake Ontario combined.

The United States and China have the largest amounts of impounded water. More than 70,000 dams in the U.S. capture and store half the annual river flow of the entire country. China has the most irrigated land of any nation in the world to feed its growing population, which might account for its many killing floods. China's water is provided by some 100,000 dams and reservoirs, which have a total storage capacity of about 100 cubic miles of water.

Interestingly, people have stored so much water in artificial reservoirs in recent decades that the added weight on the continents has subtly altered Earth's rotation, slightly shortening the length of day. Over the last half century, humans have pooled roughly 10 trillion tons of water in reservoirs, most of which are in the Northern Hemisphere. The process has shifted water from the oceans to the continents, tending to reduce mass around Earth's equator and increase it in the northern part of the globe. The shift in mass has speeded up the planet's spin by placing water closer to the axis of rotation, similar to the way a skater brings the arms closer to the body to spin faster.

Flood prevention requires engineering structures such as artificial levees and flood walls that serve as barriers against high water, building reservoirs that store excess water that is later released at safe rates, increasing the channel size to move water quickly off the land, and diverting channels to route floodwaters around areas requiring protection. The best method to minimize flood damage in urban areas is floodplain regulation along with barriers, reservoirs, and channel improvements in flood-prone areas that are already developed.

During a flood, natural levees build up on the banks of a river. When floodwaters overtop a river channel and flow onto the adjacent floodplain, the velocity quickly diminishes, causing deposition near riverbanks. The riverbanks provide a variety of environments for plant life, which helps stabilize them. The inflow of nutrients and sediments; changing water levels over the seasons, which can create different biologic niches; and waterborne dispersal of seeds contribute to a rich diversity of species occupying riverbanks.

The purpose of levees is to keep a river within its banks during normal flow. However, at flood stage, the valley floor is often lower than the river level, inundating the land when floodwaters crest over the levee top. During the 1993 Midwestern floods, perhaps the worst of the 20th century, levee breaks deposited several feet of sand onto farmlands along the swelling rivers. Artificial levees continue to break during major floods, compounding death and destruction.

Structures such as artificial levees tend to aggravate floods by forcing rivers into narrow channels instead of allowing the floodwaters to drain naturally onto the floodplains, where the power of the floods is dissipated. The effects of flooding are greatly reduced by developing a flood system based on the addition of natural wetlands that absorb excess floodwaters with less river engineering and levee construction. Wetlands can have a marked impact on the containment of most floods but are less useful during great floods, especially if heavy rains have already fully saturated the ground.

Urban areas subject to flooding should further reduce development on floodplains that require new barriers, which restrict river flow during a flood. The most practical solution is a combination of floodplain regulations and barriers that results in less physical modification of the river system. Reasonable floodplain zoning might require fewer flood prevention methods than the total absence of floodplain regulations.

Floodplain regulations are designed to obtain the most beneficial use of floodplains while minimizing flood damage and the cost of flood protection. The first step in floodplain regulations is flood hazard mapping, which provides floodplain information for land use planning. The maps delineate past floods and help derive regulations for floodplain development. These controls are a compromise between the indiscriminate use of floodplains, resulting in the destruction of property and loss of life, rather than the complete abandonment of floodplains, thereby surrendering a valuable natural resource. By recognizing the dangers of flooding and flood preparation, people can safely use what nature has reserved for excess water during a flood.

After discussing hydrology, river flow, and flooding, the next chapter shows what happens to river-borne sediments once they reach the ocean and how the sea reclaims the land.

5

COASTAL PROCESSES

SEACOASTS AND ESTUARIES

This chapter examines coastal features and the effects coastal processes have on people living near the shore. Earth is constantly evolving, with complex activities including running water and moving waves. The shifting of sediments on the land surface and the accumulation of deposits on the bottom of the ocean are continuously changing the face of the planet.

The coastal regions of the world vary substantially in topography, climate, and vegetation. Seacoasts are areas where continental and oceanic processes converge to produce a rapidly evolving landscape. Often, human intervention is necessary to reduce the effects of naturally occurring erosional events. These futile efforts eventually fail, as waves relentlessly batter the shoreline (Fig. 75).

TIDAL BASINS

Tides are caused by the pull of gravity on the ocean from the Moon and Sun and have a large influence on coastal regions (Fig. 76). If no continents

Figure 75 *Development along the shorefront Ocean City, Maryland. The distances between these buildings and the shoreline leave little room for natural processes during storms.*

(Photo by R. Dolan, courtesy USGS)

impeded the motion of the tides, all coasts would have two high tides and two low tides of nearly equal magnitude and duration each day. These are called semidiurnal tides and occur along the Atlantic coasts of North America and Europe, for example.

Other areas have different tidal patterns. The wave of the tide is reflected off and broken up by the continents. It forms a complicated series of crests and troughs thousands of miles apart. Moreover, in some regions, the tides are coupled with the motion of large, nearby bodies of water. As a result, some places, such as along the coast of the Gulf of Mexico, have only one tide a day called a diurnal tide.

Mixed tides are a combination of semidiurnal and diurnal tides such as those that occur along the Pacific coast of North America. They display a diurnal inequality with a higher-high tide, a lower-high tide, a higher-low tide, and a lower-low tide each day. Some deep-draft ships on the U.S. West Coast must often wait until the higher of the two high tides comes in before departing. A few places, such as Tahiti, have virtually no tide because they lie on a node, a stationary point about which the standing wave of the tide oscillates.

High tides that generally exceed 12 feet are called megatides. They arise in gulfs and embayments along the coast in many parts of the world. Megatides depend on the shape of the bays and estuaries, which channel the wavelike progression of the tides and increase their amplitudes. Many locations with extremely high tides also experience strong tidal currents.

Figure 76 *Beach rock formed in calcareous beach near Vega Alta, Puerto Rico.*

(Photo by W. H. Monroe, courtesy USGS)

A tidal basin near the mouth of a river can resonate with the incoming tide. The water at one end of the basin is high at the beginning, low in the middle, and high again at the end of the tidal period. The incoming tide sets the water in the basin oscillating, with water sloshing back and forth. The motion of the tide coming in toward the mouth of the river and the motion of the oscillation are synchronized. Thus, the oscillation reinforces the tide in the bay and makes the high tide higher and the low tide lower than it would be otherwise.

A special feature of this type of oscillation within a tidal basin is a tidal bore. It is a solitary wave that carries a tide upstream usually during a new or full moon. One of the largest tidal bores sweeps up the Amazon River and can rise as high as 25 feet, span several miles across, and reach 500 miles upstream. Although any body of water with high tides can generate a tidal bore, about half the known tidal bores are associated with resonance in a tidal basin.

Therefore, the tides and their resonance with the oscillation in a tidal basin provide the energy for the tidal bore.

The incoming tide arrives in a tidal basin as rapidly moving waves with long wavelengths. As the waves enter the basin, they are confined at both the sides and the bottom by the narrowing estuary. Because of this funneling action, the height of the waves increases. As the tidal bore travels upstream, it must move faster than the river current, or otherwise it is swept downstream and out to sea.

COASTAL EROSION

Steep waves that accompany storms at sea cause serious coastal erosion. The constant pounding of the surf also erodes most defenses against the rising sea. Upward of 90 percent of America's once-sandy beaches are sinking beneath the waves. Barrier islands and sandbars running along the Atlantic coast and the Gulf coast of Texas are rapidly receding. Sea cliffs are eroding back several feet a year. Eroding sea cliffs often destroy expensive homes, whose foundations are undercut by pounding waves. In California, sea cliff erosion causes huge chunks of land to fall into the sea (Fig. 77).

Figure 77 *Devils Slide caused by storm waves that continue to erode the base of the cliff, San Mateo County, California.*

(Photo by R. D. Brown, courtesy of USGS)

Half the south shore of Long Island, New York, is considered a high-risk zone for development, with the sea reclaiming some locations at a rate of up to 6 feet per year. The barrier island from Cape Henry, Virginia, to Cape Hatteras, North Carolina, has narrowed on both the seaward and landward sides (Fig. 78). The rest of the North Carolina coast is rapidly retreating from 3 to 6 feet annually, and much of the eastern coast of Texas is vanishing as well.

The pounding of the surf against the shore during severe storms is a vivid expression of the power of wave erosion. Steep waves accompanying storms at sea seriously erode sand dunes and sea cliffs. The erosion of sea cliffs and dunes that mark the coastline causes the shore to retreat a considerable distance. Most defenses erected to halt beach erosion frequently end in defeat, as the sea continues to batter the shoreline.

Waves erode by impact and pressure, by abrasion, and by solution. Wave impact can dislodge and transport large fragments. Waves running up onto shore and returning to sea move sand and pebbles back and forth, abrading sediments while simultaneously carrying them seaward. Therefore, wave erosion is similar to river erosion. Most beach material originates by wave erosion and river deposition.

Waves breaking on a coastline develop sea cliffs by undercutting the bedrock. Coastal slides occur when wave action undercuts a sea cliff, which

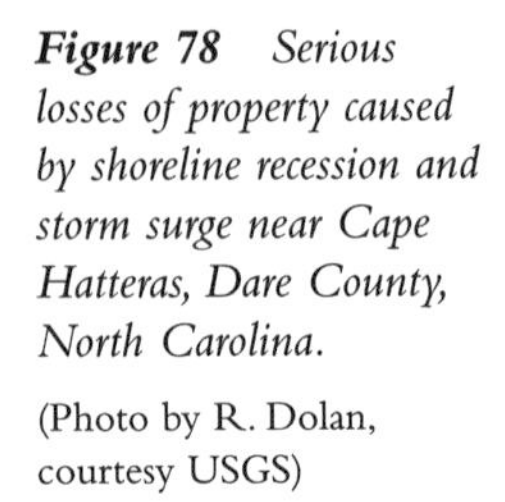

Figure 78 *Serious losses of property caused by shoreline recession and storm surge near Cape Hatteras, Dare County, North Carolina.*

(Photo by R. Dolan, courtesy USGS)

falls into the ocean. Sea cliff retreat results from marine and nonmarine agents, including wave attack, wind-driven salt spray, and mineral solution. The nonmarine agents responsible for sea cliff erosion include chemical and mechanical processes, surface drainage water, and rainfall. Mechanical erosion processes rely on cycles of freezing and thawing of water in crevasses, forcing apart fractures, which further weakens the rock.

Weathering breaks down rocks or causes the outer layers to shed by a process known as spalling. Animal trails that weaken soft rock and burrows that intersect cracks in the soil also erode sea cliffs. The sea cliff further erodes by surface water runoff and wind-driven rain. Excessive rainfall along the coast can lubricate sediments, enabling huge blocks to slide into the sea. Water running over the cliff edge and wind-driven rain produce the fluting often exposed on cliff faces.

Groundwater seeping from a sea cliff can form indentations on the cliff face, which undermines and weakens the overlying strata. The addition of water also increases pore pressure between grains within sediments, reducing the shear (plane contact) strength that holds the rock layers together. If bedding planes, fractures, or jointing dip seaward, water moving along these areas of weakness can induce rock slides. Such slides have excavated large valleys on the windward parts of the Hawaiian Islands, where powerful springs emerge from porous lava flows.

Direct wave attack at the base of a sea cliff quarries out weak beds and undercuts the cliff, causing the overlying unsupported material to collapse onto the beach. Waves also work along joint or fault planes to loosen blocks of rock or soil. In addition, winds carrying salt spray from breaking waves drive it against the sea cliff. Porous sedimentary rocks absorb the salty water, which evaporates, forming salt crystals whose growth weakens rocks. The surface of the cliff slowly flakes off and falls to the beach below, where the material landing at the base of the sea cliff piles up into a talus cone.

Limestone cliffs erode by chemical processes that dissolve soluble minerals in the rocks. Limestone erosion is common on coral islands in the South Pacific and on the limestone coasts of the Mediterranean and Adriatic Seas. Seawater dissolves the lime cement in sediments, carving deep notches in the sea cliffs. Chemical erosion also removes cementing agents, causing the sediment grains to separate.

Coastal erosion rates vary with the geography of the shoreline and the prevailing wind and tides. Beach erosion (Fig. 79) is controlled by the strength of beach dunes or sea cliffs, the intensity and frequency of coastal storms, and the exposure of the coast. Shoreline retreat is blamed on sea level rise and by long-term changes in the size and direction of waves striking the coast. The rate of coastal retreat varies with the geography of the shoreline and the prevailing wind and tides. The continuous pounding of the surf also tears down most man-made defenses against the raging sea.

The prevention of beach erosion is often thwarted because the waves constantly batter and erode structures built to keep out the sea. As a result, the methods developers use to stabilize the seashores are destroying the very beaches upon which they intend to build. Often, beach erosion is aggravated by the structures engineers erect to stabilize the shoreline (Fig. 80). Jetties and seawalls built to halt the tides tend to increase erosion. Jetties cut off the natural supply of sand to beaches, and seawalls increase erosion by bouncing waves back instead of absorbing their energy. The rebounding waves carry sand out to sea, undermining the beach and destroying the shorefront property.

Coastal residents often build expensive seawalls to protect houses on eroding bluffs overlooking the sea. Unfortunately, these structures tend to hasten the erosion of beach sands in front of the walls. Beaches forward of the seawalls often lose sand during certain seasons, while waves return beach sands at other times. In effect, the seawalls are saving the bluff at the detriment of the beaches. Barriers erected at the bottom of sea cliffs might deter wave erosion but do not affect sea spray and other erosion processes.

Figure 79 *Beach wave erosion at Grand Isle, Louisiana, on August 30, 1985, from Hurricanes Danny and Elana.*

(Photo courtesy U.S. Army Corps of Engineers)

Figure 80 *A groin field along Fenwick Island, Worcester County, Maryland.*

(Photo by R. Dolan, courtesy USGS)

COASTAL DEPOSITION

Rivers and streams deliver to the coast a heavy load of sediment washed off the continents (Fig. 81). River-borne sediments entering the ocean settle onto the continental shelf, which extends up to 100 miles or more and reaches a depth of roughly 600 feet. In most places, the continental shelf is nearly flat lying, with an average slope of about 10 feet per mile, comparable to the slopes

Figure 81 *The delta of Camerai Creek at the north end of Waterton Lake, Alberta, Canada.*
(Photo by C. D. Walcott, courtesy USGS)

of many coastal regions. Indeed, during the ice ages, these areas were the seacoasts of the world, when the ocean dropped several hundred feet.

Marine sediments comprise quartz grains about the size of beach sands. Many marine sandstone formations such as those exposed in the American West were deposited along the shores of ancient inland seas. When reaching the coast, the river-borne sediments settle out of suspension according to grain size. The coarse-grained sediments deposit near the turbulent shore, while the fine-grained sediments settle in calmer waters farther out to sea. As the shoreline advances seaward from a buildup of coastal sediments or a falling sea level, finer sediments are overlain by progressively coarser ones. As the shoreline recedes landward from a lowering of the land surface or a rising sea level, coarser sediments are overlain by progressively finer ones. The differing sedimentation rates as the sea transgresses and regresses result in a recurring sequence of sands, silts, and muds.

The overlying sedimentary layers pressing down onto the lower strata and cementing agents such as calcite and silica transform the sediments into solid rock, providing a geologic column of alternating beds of limestones, shales, siltstones, and sandstones. Abrasion eventually grinds all rocks down to clay-sized particles, the most abundant sediments. The minute particles sink slowly, settling out in calm, deep waters far from shore.

Sediment layers vary in thickness according to the sedimentary environment at the time they were laid down. Bedding planes mark where one type of deposit ends and another begins. Thus, thick sandstone beds might be interspersed with thin beds of shale and siltstone. This indicates periods when coarse sediments were deposited punctuated by periods when fine sediments were laid down, as the shoreline progressed and receded.

Graded bedding results from the varying of particle size in a sedimentary bed from coarse at the bottom to fine at the top. This indicates rapid deposition of sediments of differing sizes by fast-flowing streams emptying into the sea. The largest particles settle out first and are covered by progressively finer material due to the difference in settling rates. Beds also grade laterally, resulting in a horizontal gradation of sediments from coarse to fine.

Sedimentary beds also vary in color, which helps identify the type of depositional environment. Generally, sediments tinted various shades of red or brown indicate a terrestrial source, whereas green- or gray-colored sediments suggest a marine origin. The size of individual particles influences the color intensity. Usually, darker colored sediments indicate finer grains.

Limestones deposited onto the shallow floors of oceans or large lakes are among the most common rocks. They make up about 10 percent of the land surface. They are composed of calcium carbonate mostly derived from biological activity as evidenced by abundant fossils of marine life in limestone beds. Chalk is a soft, porous carbonate rock. One of the largest deposits of chalk is the cliffs of Dorset, England, where poor consolidation of the strata results in severe erosion during coastal storms. Many limestone formations comprised ancient coral reefs exposed on dry land.

Coral reefs are important to the coastal geology and play a major role in altering continental shorelines. The reefs are limited to clear, warm, sunlit tropical waters in the Indo-Pacific and the western Atlantic (Fig. 82). About 270,000 square miles of coral reefs are estimated to exist in the world's oceans. Over geologic time, corals and other organisms living on the reefs have built massive formations of limestone.

A typical reef consists of fine, sandy detritus, stabilized by plants and animals anchored to the surface. The corals' ability to build wave-resistant structures thereby encouraged tropical plant and animal communities to thrive on the reefs, which are thought to house one in every four marine species. Hundreds of species of encrusting organisms such as barnacles thrive on the coral reef. Smaller, more fragile corals and large communities of green and red calcareous algae live on the coral framework.

The coral rampart reaches almost to the water's surface. It consists of large rounded coral heads and a variety of branching corals. The fore reef is seaward of the reef crest, where corals blanket nearly the entire seafloor. In deeper waters, many corals grow in flat, thin sheets to maximize their light-gathering area. In other parts of the reef, the corals form large buttresses separated by narrow sandy channels composed of calcareous debris from dead corals, calcareous algae, and other organisms living on the coral. The channels resemble narrow, winding canyons with vertical walls of solid corals. They dissipate wave energy and allow the free flow of sediments, which prevents the coral from choking on the debris. Below the fore reef is a coral terrace, followed by a sandy slope with isolated coral pinnacles, then another terrace, and finally a nearly vertical drop into the dark abyss.

Fringing reefs (Fig. 83) grow in shallow seas. They hug the coastline or are separated from the shore by a narrow stretch of water. Barrier reefs also parallel the coast but lie farther out to sea. They are much larger and extend for longer distances. The best example is the Great Barrier Reef, a chain of more than 2,500 coral reefs and small islands off the northeastern coast of Australia. It forms an undersea embankment more than 1,200 miles long, up to 90 miles wide, and as much as 400 feet high. The reef is the largest feature built by living organisms and harbors about 400 species of coral.

The second largest barrier reef is the Belize Barrier Reef complex on the Caribbean coast of South America. It is the most luxuriant array of reefs in the Western Hemisphere. Extensive reefs also rim the Bahama Banks archipelago. A small reef that fringes Costa Rica is in danger from pollution from pesticides and soil runoff. Other reefs throughout the world are similarly affected.

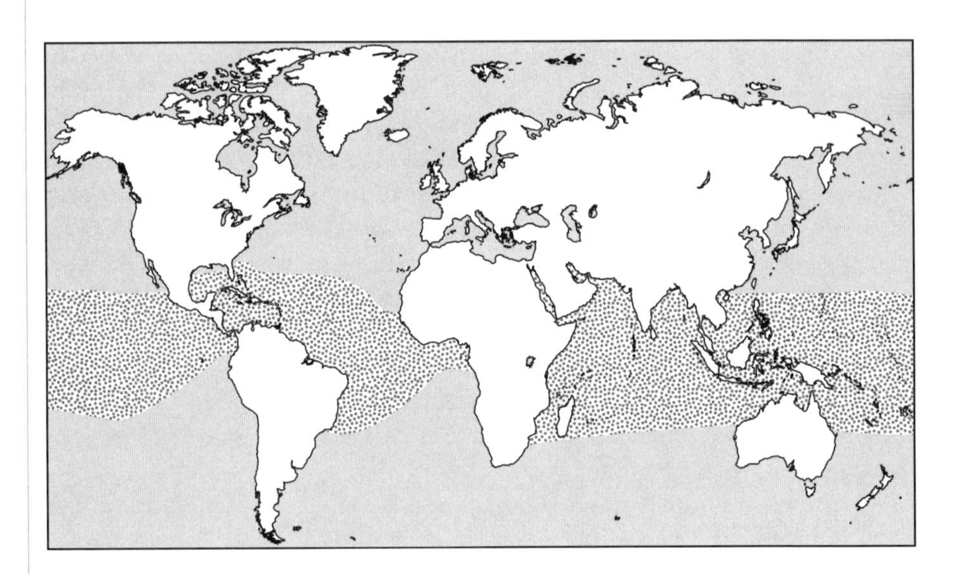

Figure 82 *Worldwide belt of coral reefs.*

Figure 83 *A fringing coral reef on the south coast of Puerto Rico.*

(Photo by C. A. Kaye, courtesy USGS)

COASTAL SUBSIDENCE

Subsidence is the downward settling of earth materials with little horizontal movement. Coastal areas often subside during large earthquakes when one block of crust drops below another. During an earthquake, vegetated lowlands elevated to avoid inundation by the sea submerge regularly and become barren tidal mudflats. Between quakes, sediments fill the tidal flats, raising them to the level where vegetation can grow once again. Therefore, repeated earthquakes produce alternating layers of lowland soil and tidal flat mud (Fig. 84).

During the 1964 Alaskan earthquake, more than 70,000 square miles of land tilted downward, causing extensive flooding in coastal areas of southern Alaska. The earthquake produced submarine flow failures that destroyed many seaport facilities. The flow failures also generated large tsunamis that overran coastal areas, causing additional damage and casualties. Coastal regions in Japan are particularly susceptible to subsidence. Had the January 17, 1995, Kobe earthquake of 7.2 magnitude struck Tokyo instead, more than half the city would have sunk beneath the waves.

Subsidence occurs when fluids pumped from subterranean sediments result in compaction. The overdrawing of groundwater in the northeastern section of Tokyo, Japan, has caused the land to sink at a rate of about half a

Figure 84 *Secondary cracks in the tidal flat at the head of Bolinas Lagoon, Marin County, California, from the 1906 California earthquake.*

(Photo by G. K. Gilbert, courtesy USGS)

foot a year over an area of about 40 square miles. About 15 square miles sank below sea level, requiring the construction of dikes to keep out the sea from certain sections of the city during a typhoon or an earthquake.

Some of the most dramatic examples of subsidence are along the seacoasts (Fig. 85). Coastal cities subside due to a combination of rising sea level and withdrawal of groundwater, causing the aquifer to compact. Subsidence in some coastal areas has increased susceptibility to flooding during earthquakes or severe coastal storms. Parts of Niigata, Japan, sank below sea level during the extraction of water-saturated natural gas, requiring the construction of dikes to keep out the sea. During the June 16, 1964, earthquake, the dikes were breached with seawater. The city subsided 1 foot or more, causing serious flooding in the area of subsidence. A tsunami generated by the earthquake also damaged the harbor area.

The Nile Delta of Egypt is heavily irrigated and supports 50 million people in a 7,500 square mile area. Port Said on the northeast coast of the delta sits at the northern entrance to the Suez Canal. The region overlies a large depression filled with 160 feet of mud, indicating that part of the delta is slowly dropping into the sea. Over the last 8,500 years, this portion of the

Figure 85 *Submergent coastline north of Portland, Lincoln County, Maine.*

(Photo by J. R. Balsley, courtesy USGS)

fan-shaped delta has been lowering by less than a quarter inch per year. More recently, though, the yearly combined subsidence and sea level rise have greatly exceeded this amount, which could place major portions of the city underwater. Moreover, as the land subsides, seawater infiltrates into the groundwater system, rendering it useless.

Venice, Italy, is drowning because of a combination of rising sea levels and subsidence. The city is most unusual because it is built right at the water's edge, where buildings rest half on sea and half on land. Venice has been battling rising waters since the 5th century, with the pace of change accelerating markedly over the last 100 years. The city is built on soft, compact sediments and is slowly sinking under its own weight. Venice has sunk more than 6 feet since its founding, forcing residents to fill in the lagoon with sand to stay above water. The city regularly floods during high tides, heavy spring runoffs, and storm surges.

COASTAL INUNDATION

Sea levels have always been changing throughout geologic history. More than 30 rises and falls of global sea levels occurred between 6 and 2 million years ago. At its highest point between 5 and 3 million years ago, the global sea level rose about 140 feet higher than it is today. Between 3 and 2 million years ago, the sea level dropped at least 65 feet lower than at present due to growing glaciers at the poles. During the ice ages, sea levels dropped as much as 400 feet at the peak of glaciation. The global sea level steadied about 6,000 years ago after rising rapidly for thousands of years following the melting of the great glaciers of the last ice age.

Civilizations have endured changing sea levels for centuries (Table 6). If the ocean continues to rise, the Dutch who reclaimed their land from the sea would find a large portion of their country lying underwater. Many islands would drown or become mere skeletons of their former selves with only their mountainous backbones showing above the water. Half the scattered islands of the Republic of Maldives southwest of India would be lost. Much of Bangladesh would also drown, a particularly distressing situation since the region heavily floods during typhoons. Most of the major cities of the world, because they are located on seacoasts or along inland waterways, would be inundated by the sea with only the tallest skyscrapers poking above the waterline. Coastal cities would have to rebuild farther inland or construct protective seawalls to hold back the waters.

Over the last century, the global sea level appears to have risen upward of 9 inches due mainly to the melting of the Antarctic and Greenland ice sheets. The present rate of sea level rise is several times faster than 40 years ago,

TABLE 6 MAJOR CHANGES IN SEA LEVEL

Date	Sea level	Historical event
2200 B.C.	Low	
1600 B.C.	High	Coastal forest in Britain inundated by the sea.
1400 B.C.	Low	
1200 B.C.	High	Egyptian ruler Ramses II builds first Suez canal.
500 B.C.	Low	Many Greek and Phoenician ports built around this time are now under water.
200 B.C.	Normal	
A.D. 100	High	Port constructed well inland of present-day Haifa, Israel.
A.D. 200	Normal	
A.D. 400	High	
A.D. 600	Low	Port of Ravenna, Italy becomes landlocked. Venice is built and is presently being inundated by the Adriatic Sea.
A.D. 800	High	
A.D. 1200	Low	Europeans exploit low-lying salt marshes.
A.D. 1400	High	Extensive flooding in low countries along the North Sea. The Dutch begin building dikes.

amounting to about an inch every five years. Greenland holds about 6 percent of the world's freshwater in its ice sheet (Fig. 86). An apparent warming climate is melting more than 50 billion tons of water a year from the Greenland ice sheet, amounting to more than 11 cubic miles of ice annually. In addition, higher global temperatures could influence Arctic storms, increasing the snowfall in Greenland 4 percent with every 1 degree Celsius rise in temperature.

The melting of the Greenland ice and the calving of icebergs from glaciers entering the sea (Fig. 87) is responsible for about 7 percent of the yearly rise in global sea level. The Greenland ice sheet is undergoing significant thinning of the southern and southeastern margins, in places as much as 7 feet a year. Furthermore, Greenland glaciers are moving more rapidly to the sea, possibly caused by meltwater at the base of the glaciers that helps lubricate the downhill slide of the ice streams. The grounding line is the point where the glacier reaches the ocean and the ice lifts off the bedrocks and floats as an iceberg. In an average year, some 500 icebergs spawn from western Greenland and drift down the Labrador coast, where they become shipping hazards. In 1912, the S.S. *Titanic* was sunk by such an iceberg.

A sustained warmer climate could cause the polar ice caps to melt. The melting is increasing the risk of coastal flooding around the world during high

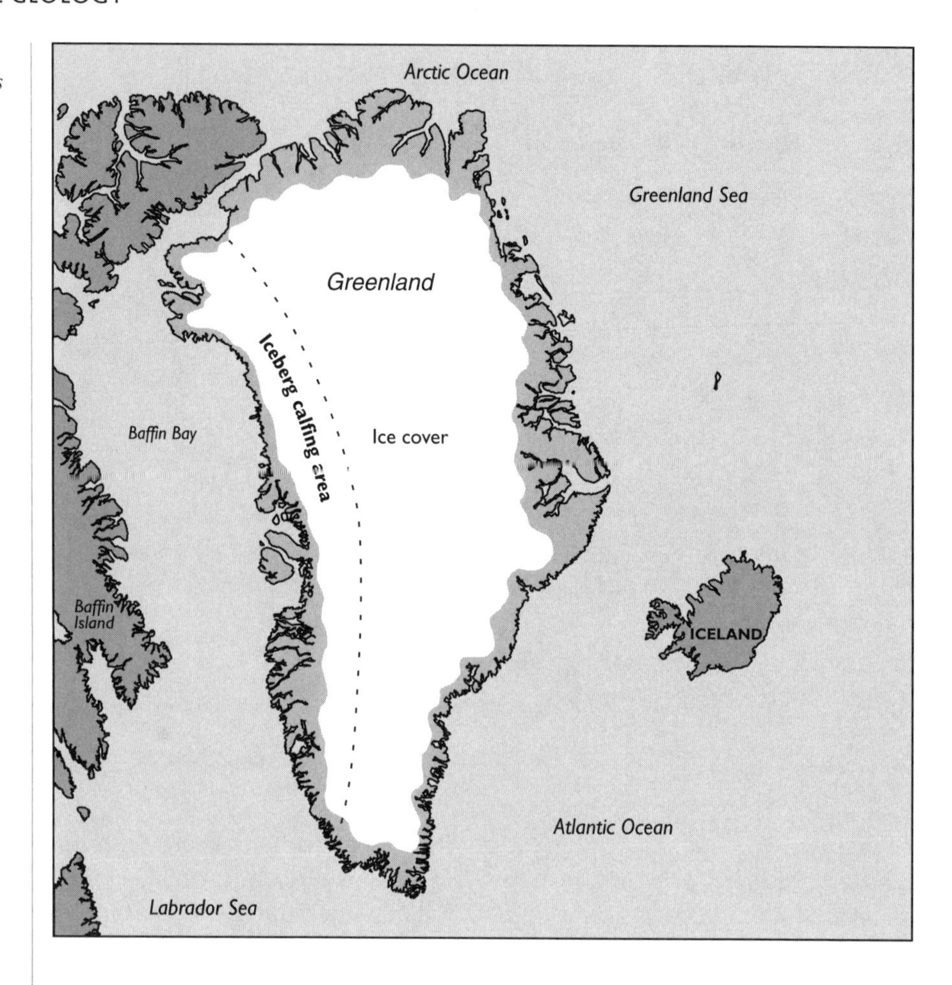

Figure 86 *The Greenland ice sheet holds much of the world's ice.*

tides and storms. The calving of large numbers of icebergs from glaciers entering the ocean could substantially raise sea levels, thereby drowning coastal regions. The additional freshwater in the North Atlantic could also affect the flow of the Gulf Stream, causing Europe to freeze while the rest of the world continues to warm.

At the present rate of melting, the sea could rise 1 foot or more by the middle of this century, comparable with the melting rate of the continental glaciers at the end of the last ice age. The rapid deglaciation between 16,000 and 6,000 years ago, when torrents of meltwater entered the ocean, raised the sea level on a yearly basis only a few times greater than it is rising today.

As global temperatures increase, coastal regions where half the people of the world live would feel the adverse effects of rising sea levels due to melting ice caps and thermal expansion of the ocean. In areas such as Louisiana, the level of the sea has risen upward of 3 feet per century. The thermal expansion of the ocean has also raised the sea level about 2 inches. Surface waters off the California coast have warmed nearly 1 degree Celsius over the past half century, causing the water to expand and raise the sea about 1.5 inches.

Higher sea levels are also caused in part by sinking coastal lands due to the increased weight of seawater pressing down onto the continental shelf. In addition, sea level measurements are affected by the rising and sinking of the land surface due to plate tectonics and the rebounding of the continents after the glacial melting at the end of the last ice age.

If all the polar ice melted, the additional seawater would move the shoreline up to 70 miles inland in most places. Low-lying river deltas that feed much of the world's population would be inundated by the rising waters. The inundation would radically alter the shapes of the continents. The receding shore would result in the loss of large tracks of coastal land along with shallow barrier islands. Estuaries, where marine species hatch their young, would be destroyed. All of Florida along with south Georgia and the eastern Carolinas would vanish. The gulf coastal plain of Mississippi, Louisiana, east Texas, and major parts of Alabama and Arkansas would virtually disappear. Much of the Isthmus separating North and South America would sink out of sight.

Figure 87 *A Coast Guard icebreaker approaches a large tabular iceberg in Melville Bay, Greenland.*

(Photo courtesy U.S. Coast Guard)

If the present melting continues, the sea could rise a foot or more by the middle of the century. For every foot of sea level rise, 100 to 1,000 feet of shoreline would be inundated, depending on the slope of the coast. Just a 3-foot rise could flood about 7,000 square miles of coastal land in the United States, including most of the Mississippi Delta, possibly reaching the outskirts of New Orleans.

The sea level is rising as much as 10 times faster than it did a century ago, amounting to about a quarter inch per year. Most of the increase appears to result from melting ice caps, particularly in West Antarctica and Greenland. Most of the ice flowing into the sea from the Antarctic ice sheet discharges from a small number of fast-moving ice streams and outlet glaciers. In addition, more icebergs are calving off glaciers entering the sea. They appear to be getting larger as well, threatening the stability of the ice sheets. The number of extremely large icebergs has also increased dramatically. Much of this instability is blamed on global warming.

Alpine glaciers, which contain substantial quantities of ice (Fig. 88), are melting as well, possibly due to a warmer climate. Some areas such as the European Alps might have lost more than half their cover of ice. Moreover, the rate of loss appears to be accelerating. Tropical glaciers such as those in the high mountains of Indonesia have receded at a rate of 150 feet per year over

Figure 88 *Chocolate Glacier on the eastern side of Glacier Peak was very active and advanced between 1950 and 1968, Snohomish County, Washington.*

(Photo by A. Post, courtesy USGS)

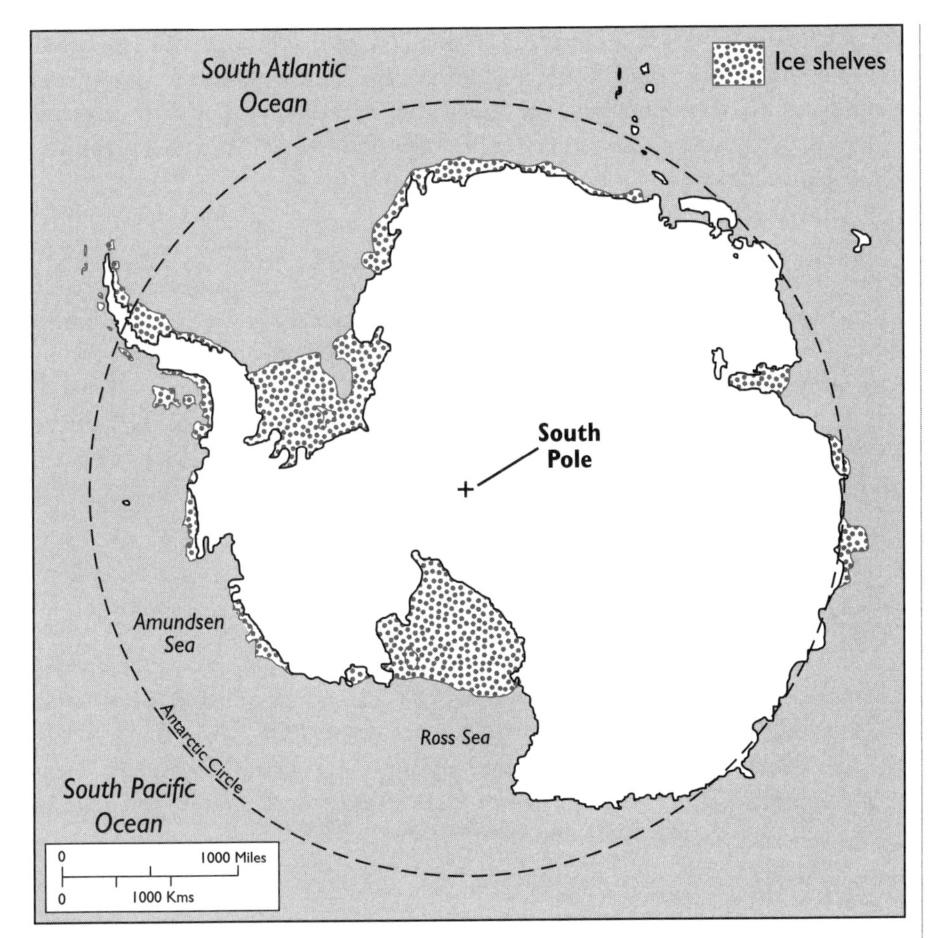

Figure 89 The ice shelves in Antarctica shown in stippled areas.

the last two decades. At the present rise in temperature and rate of retreat, the glaciers are likely to disappear altogether.

Sea ice covers most of the Arctic Ocean and forms a frozen band around Antarctica during the winter season in each hemisphere. These polar regions are most sensitive to global warming and experience greater atmospheric changes than other parts of the world. About half of Antarctica is bordered by ice shelves (Fig. 89). The two largest, the Ross and Filchner-Ronne, are nearly the size of Texas. The 2,600-foot-thick Filchner-Ronne ice shelf might actually thicken with global warming, which would enhance the ice-making process. Many other ice shelves could become unstable and float freely in a warmer climate. Since the 1950s, several smaller ice shelves have disintegrated. Today, some large shelves are starting to retreat.

A warm interlude between ice ages around 400,000 years ago, known as stage II, was a 30,000-year-period of global warming that eclipsed that of today. During this time, the melting of the ice caps caused the sea level to rise about 60 feet higher than at present. Most of the high seas were caused by the melting of the West Antarctica ice shelves, leaving open ocean in their place. The rest came from the melting of the stable East Antarctica ice cap and the Greenland ice sheet.

If average global temperatures continue to rise, this interglacial could become equally as warm if not warmer than stage II. The warmer climate could induce an instability in the West Antarctic ice sheet, causing it to surge into the sea. This rapid flow of ice into the ocean could raise sea level up to 20 feet or more, inundate the continents several miles inland, and flood valuable property. In the United States alone, a full quarter of the population would find itself underwater, mostly along the East and Gulf coasts. If all the ice on Antarctica, which holds 90 percent of the world's total, were to melt, enough water would be dumped into the ocean to raise global sea levels nearly 200 feet.

COASTAL FLOODS

Torrential downpours and tidal floods from hurricanes and typhoons cause more damage and take more lives than other forms of flooding (Fig. 90). By their very nature, tropical storms drop huge amounts of rainfall over large areas often within a day or so. The deluge causes widespread flooding in natural

Figure 90 *Floodwaters during the July 21, 1974, Typhoon Ivy at the Subic Bay Naval Station, Philippines.*

(Photo by B. A. Richards, courtesy U.S. Navy)

Figure 91 *Overwash and storm surge penetration near Cape Hatteras, North Carolina, in 1962.*

(Photo by R. Dolan, courtesy USGS)

drainage areas, where streams cannot cope with the excess water formed by the onrush of heavy rains.

Tidal floods are overflows on coastal lands, such as bars, spits, and deltas, which are affected by the coastal current and occupy the same protective position relative to the sea that floodplains do with rivers. Most of the severe tidal floods are caused by tidal waves generated by high winds from hurricanes superimposed onto the regular cycle of tides resulting in a storm surge (Fig. 91). Hurricane winds often produce wave heights several feet higher than the maximum level of the prevailing high tide. Tidal floods also result from the combination of waves generated by hurricane winds and flood runoff from heavy rains that accompany the storms.

The flooding extends over large distances along a coastline. The duration is usually short, depending on the elevation of the tide, which usually rises and falls twice daily. When the tide is in, other factors that produce high waves can raise the maximum level of the prevailing high tide. Hurricanes are the primary sources of extreme winds and high waves. Every year, severe storms enter the American mainland. They cause a tremendous amount of damage and flooding as well as severe beach erosion that continues to move the coastline landward.

The deadliest natural disaster in American history devastated the Galveston, Texas, area on September 8, 1900. The city was literally built on an island of sand with an average elevation of only a few feet about sea level. A hurricane with wind speeds of more than 110 miles per hour sent a storm surge through town, crumbling buildings and sending people into the surging waters. When calm finally returned, 10,000 to 12,000 people were counted among the dead.

The Bay of Bengal, Bangladesh, with a population of more than 100 million crammed into a country no larger than Wisconsin, is frequently swamped by typhoons originating on the Indian Ocean. On May 24, 1985, a powerful typhoon surged up the bay, accompanied by up to 50-foot-tall waves that swept over a cluster of islands. The storm ravaged 3,000 square miles of valuable cropland and ruined vital fishing grounds. When the storm ended and the waters receded, upward of 100,000 people were found dead, and 250,000 were left homeless.

WAVE ACTION

Large waves breaking on a seacoast are striking examples of the sizable amount of energy ocean waves generate. The intertidal zones of rocky-weather coasts receive much more energy per unit area from waves than they do from the Sun. The waves are created by strong winds generated by distant storms blowing across large areas of the open ocean. Local storms near the coasts provide the strongest waves, especially when superimposed onto rising tides.

Most waves are generated by large storms at sea as strong winds blow across the surface of the ocean. Waves breaking on the coast dissipate energy and generate alongshore currents that transport, sand along the beach. High waves during coastal storms cause most beach erosion, a serious problem in areas where the shoreline is steadily receding. Sudden barometric pressure changes on large lakes or bays can cause water to slosh back and forth, producing waves called seiches. They are common on Lake Michigan and on occasions can be quite destructive.

The most dramatic storm surges are produce by hurricanes, which are responsible for destroying entire beaches. As the wave approaches shore, it touches bottom and slows. The shoaling of the wave in shallow water distorts its shape, making it break upon the beach. The breaking wave dissipates its energy along the coast and erodes the shoreline.

Wave energy reflecting off steep beaches or seawalls forms sandbars. When waves approach the shore at an angle to the beach, the wave crests bend by refraction. As waves pass the end of a point of land or the tip of a breakwater, a circular wave pattern generates behind the breakwater. When the refracted waves intersect other incoming waves, they increase the wave height.

Swells reaching a coast produce various types of breakers (Fig. 92), depending on the wave steepness and bottom slope conditions near the beach. If the bottom slope is relatively flat, the wave forms a spilling breaker, the most common type. It is an oversteepened wave that starts to break at the crest and continues breaking as it heads toward the beach. If the bottom slope increase to about 10 degrees, the wave forms a plunging breaker. The crest curls over, creating a tube of water. As the wave breaks, the tube moves toward the bottom and stirs up sediments. Plunging waves are the most dramatic breakers and do the most beach damage because the energy concentrates at the point where the wave breaks.

If the bottom slope steepens to about 15 degrees, the wave forms a collapsing breaker. The breaker is confined to the lower half of the wave. However, as the wave moves toward the coast, most of it reflects off the beach. On a steep bottom where the slope is greater than 15 degrees, a surging wave develops. The wave does not break but surges up the beach face and reflects

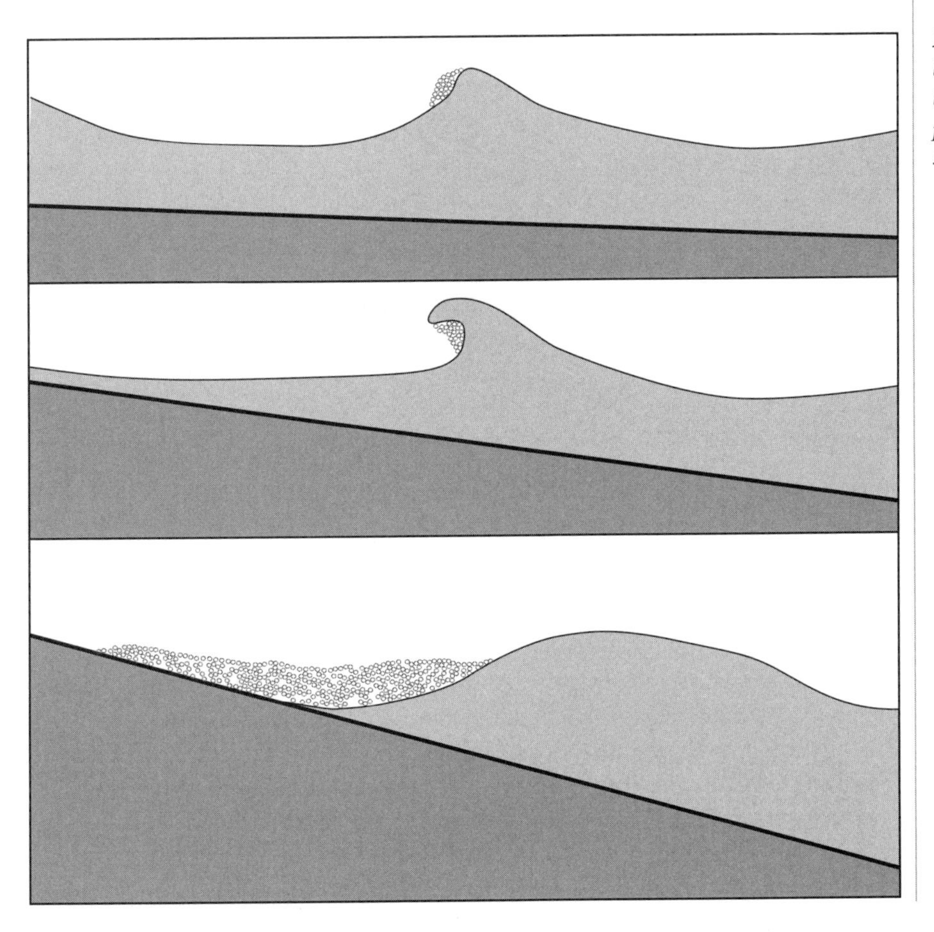

Figure 92 *Types of breakers: (from top to bottom) spilling breaker, plunging breaker, and surging breaker.*

off the coast, generating standing waves near the shore. Standing waves are important for the development of offshore structures such as bars, sand spits, beach cusps, and riptides.

TSUNAMIS

Tsunamis are the world's most destructive waves. Undersea and nearshore earthquakes produce powerful tsunamis, a Japanese word, meaning "harbor waves," so-named because of their common occurrence in this region. The vertical displacement of the ocean floor during earthquakes causes the most destructive tsunamis, whose wave energy is proportional to the seismic intensity. The earthquake sets up ripples on the ocean similar to those formed by tossing a rock into a quiet pond.

In the open ocean, the waves extend thousands of feet downward to the seafloor. Generally, the deeper the water and the longer the wave, the faster the tsunami travels. The wave crests span up to 300 miles long, but the wave height is usually less than 3 feet. The distance between crests, or the wave length, is 60 to 120 miles, giving the tsunami a very gentle slope, which allows it to pass by ships practically unnoticed. Tsunamis travel at speeds between 300 and 600 miles per hour. Upon entering shallow coastal waters, they have been known to grow into a wall of water up to 200 feet high, although most tsunamis reach a height of only a few tens of feet.

When the wave touches bottom in a harbor or narrow inlet, its speed diminishes rapidly to about 100 miles per hour. The sudden breaking action causes seawater to pile up. The wave height is magnified tremendously as waves overtake one another, decreasing the distance between them in a process called shoaling. The destructive power of the wave is immense, and the damage it causes as it crashes to shore is considerable. Large buildings are crushed with ease, and vessels are tossed up and carried well inland (Fig. 93).

The Pacific Ocean is responsible for 90 percent of all tsunamis in the world, and 85 percent of those are the products of undersea earthquakes. Between 1992 and 1996, 17 tsunami attacks around the Pacific killed some 1,700 people. The Hawaiian Islands are in the paths of many damaging tsunamis. Since 1895, 12 such waves have struck the islands. In the most destructive tsunami, 159 people died in Hilo on April 1, 1946 by killer waves generated by a powerful earthquake in the Aleutian Islands to the north.

The Good Friday earthquake on March 27, 1964, the largest recorded to hit the North American continent, devastated Anchorage, Alaska, and surrounding areas. The 9.2-magnitude quake cause destruction over an area of 50,000 square miles and was felt throughout an area of 500,000 square miles. A 30-foot-high tsunami generated by the undersea earthquake destroyed

Figure 93 *Tsunamis destruction of Kodiak Island from the March 27, 1964, Alaska earthquake.*

(Photo courtesy NOAA)

coastal villages around the Gulf of Alaska, killing 107 people. Kodiak Island was heavily damaged. Most of the fishing fleet was destroyed there when the tsunami carried many vessels inland. As a striking example of a tsunami's great power, large spruce trees were snapped off with ease by a large tsunami near Shoup Bay (Fig. 94).

Surprisingly, the San Andreas Fault, which runs under California and into the ocean, hardly causes a ripple because plates slide horizontally past each other instead of moving up and down. The sudden change in seafloor terrain triggers tsunamis when the seabed rapidly sinks or rises during an earthquake. This either lowers or rises an enormous mound of water, stretching from the seafloor to the surface. The mound of water thrust above normal sea level quickly collapses under the pull of gravity. The vast swell can cover up to 10,000 square miles, depending on the area uplifted on the ocean floor. This alternating swell and collapse spreads out in concentric rings on the surface of the ocean.

Prior to the establishment of a tsunami watch in the Pacific Ocean, people had little warning of impending disaster except for the rapid withdrawal of seawater from the shore. Residents of coastal areas frequently stricken by tsunamis heed this warning and head for higher ground. When a freak tsunami struck on the island of Madeira in the Azores after the 1755 Lisbon, Portugal,

Figure 94 *Spruce trees 2 feet in diameter were snapped off by a local wave at elevations between 88 and 101 feet near Shoup Bay, Valdez district from the March 27, 1964, Alaskan earthquake.*

(Photo by G. Plafker, courtesy USGS)

earthquake, large quantities of fish were stranded onshore as the sea suddenly retreated. Villagers, unaware of any danger, ventured out to collect this unexpected bounty, only to lose their lives when, without warning, a gigantic wave crashed down onto them.

A few minutes after the sea retreats, a tremendous surge of water pounds the shore, extending hundreds of feet inland. Often a succession of surges occurs, each followed by a rapid retreat of water back to sea. On coasts and islands where the seafloor rises gradually or where protective barrier islands exist, much of the tsunami's energy is spent before it ever reaches shore. However, on volcanic islands surrounded by very deep water or where deep submarine trenches lie immediately outside harbors, an oncoming tsunami can build to tremendous heights.

Destructive tsunamis generated by large earthquakes can travel completely across the Pacific Ocean. The great 1960 Chilean earthquake elevated a California-sized chunk of land about 30 feet and created a 35-foot tsunami that struck Hilo, Hawaii (Fig. 95), more than 5,000 miles away. The tsunami caused more than 20 million dollars in property damages and 61 deaths. The tsunami traveled an additional 5,000 miles to Japan and inflicted considerable destruction onto the coastal villages of Honshu and Okinawa, leaving 180 people dead or missing. In the Philippines, 20 people were killed. Coastal areas of New Zealand were also damaged. For several days afterward, tidal gauges in Hilo could still detect the waves as they bounced around the Pacific Basin.

The most tsunami-prone area in the world is the Pacific rim, which experiences the most earthquakes as well as the most volcanoes. A tsunami originating in Alaska could reach Hawaii in 6 hours, Japan in 9 hours, the

Figure 95 *Damage from the 1960 Hilo, Hawaii, tsunami.*
(Photo courtesy NOAA)

Philippines in 14 hours. A tsunami originating off the Chilean coast could reach Hawaii in 15 hours and Japan in 22 hours, time enough to take the necessary safety precautions that might spell the difference between life and death.

Seismic sea wave reporting stations administered by the National Weather Service are stationed in various parts of the Pacific, which is responsible for about 90 percent of all recorded tsunamis in the world. When an earthquake of 7.5 magnitude or more occurs in the Pacific area, the epicenter is plotted and the magnitude is calculated. A tsunami watch is put out to all stations in the network. The military and civilian authorities concerned are also notified. Each station in the network detects and reports the sea waves as they pass in order to monitor the progress of the tsunami. The data is used to calculate when the wave is likely to reach the many populated areas at risk around the Pacific.

Unfortunately, the unpredictable nature of tsunamis causes many false warnings that result in areas being evacuated unnecessarily or residents ignoring the warnings altogether. This happened on May 7, 1986, when a tsunami predicted for the West Coast from the 7.7-magnitude Adak earthquake in the Aleutians, for some reason, failed to arrive. People ignored a similar tsunami warning in Hilo in 1960 at the cost of their lives. Not much can be done to prevent damage from tsunamis. However, when given the advance warning time, coastal regions can be evacuated successfully with minimal loss of life.

After discussing the geologic activity of the seacoasts, the next chapter investigates the tectonic hazards resulting from earthquakes and volcanoes.

6

TECTONIC HAZARDS

EARTHQUAKES AND VOLCANOES

This chapter examines the effects of ground shaking and volcanic activity and their dangers to society. Earthquakes are extremely hazardous in fault-prone regions. In mere seconds, a major temblor can practically level an entire city. Earthquakes are by far the most powerful short-term natural forces. A sudden, massive jolt can strike without eminent warning. In the last 500 years, some 3 million people have lost their lives in earthquakes (Table 7). Their disruptive effects on sociology and the economy can be devastating, leading to starvation, disease, and other secondary reverberations. Earthquakes are considered man-made disasters because most deaths occur when buildings collapse onto people. As the world's population continues to grow near active faults (Fig. 96), catastrophes from major earthquakes could rise dramatically.

Volcanoes are natural hazards that are highly destructive to societies living within their domain. Most historic volcanic eruptions have caused fatalities (Table 8). Over the last 1,000 years, as many as 500,000 people have become victims to volcanoes. During the last century, volcanoes caused an average death toll of roughly 1,000 people a year. In the 1980s alone, when volcanic eruptions appeared to be on the rise, some 40,000 people lost their

TABLE 7 THE MOST DESTRUCTIVE EARTHQUAKES

Date (A.D.)	Region	Magnitude	Death toll
365	Eastern Mediterranean		5,000
478	Antioch, Turkey		30,000
856	Corinth, Greece		45,000
1042	Tabriz, Iran		40,000
1556	Shenshu, China		830,000
1596	Uryu-Jima, Japan		4,000
1737	Calcutta, India		300,000
1755	Lisbon, Portugal		60,000
1757	Concepción, Chile		5,000
1802	Tokyo, Japan		200,000
1811	New Madrid, Missouri		<1,000
1812	Caracas, Venezuela		10,000
1822	Valparaíso, Chile		10,000
1835	Concepción, Chile		5,000
1857	Tokyo, Japan		107,000
1866	Peru & Ecuador		25,000
1877	Ecuador		20,000
1883	Dutch Indies		36,000
1891	Mino-Owari, Japan		7,000
1902	Martinique, West Indies		40,000
1902	Guatemala		12,000
1906	San Francisco, California	8.2	3,000
1908	Messina, Sicily	7.5	73,000
1915	Italy		29,000
1920	Kansu, China	8.6	180,000
1923	Tokyo/Yokohama, Japan	8.3	143,000
1927	China	8.6	70,000
1935	Quefta, Pakistan		40,000
1939	Concepción, Chile		50,000
1939	Erzincan, Turkey	7.9	23,000
1949	Tadzhikstan		12,000

TABLE 7 *(CONTINUED)*

Date (A.D.)	Region	Magnitude	Death toll
1949	Ecuador		6,000
1953	Greece		3,000
1960	Agadir, Morocco	5.7	12,000
1960	Chile	9.5	6,000
1962	Iran		12,000
1968	Iran		12,000
1970	Peru		67,000
1972	Iran		5,500
1972	Managua, Nicaragua	6.2	12,000
1976	Guatemala	7.5	22,000
1976	Tangshan, China	7.6	240,000
1976	Turkey	7.3	4,000
1978	Eastern Iran		25,000
1980	Southern Italy		45,000
1981	Southeastern Iran		8,000
1982	Northern Yemen		3,000
1985	Mexico City, Mexico	7.8	8,000
1988	Spitak, Armenia	6.9	8,000
1990	Northern Iran		100,000
1995	Kobe, Japan	7.2	5,500
1999	Northern Turkey	7.4	17,000
2001	Western India	7.9	30,000

lives to deadly volcanoes. The increased volcano-related deaths are primarily due to the growing numbers of people living near the domain of active volcanoes and not necessarily due to more eruptions. Presently, about 600 active volcanoes that have erupted in historic times exist throughout the world, with hundreds more dormant or extinct cones that could reawaken anytime.

GROUND SHAKING

Ground shaking describes the vibration of the land during earthquakes. It is caused by surface waves and body waves, which penetrate Earth's interior. Most earthquake damage results from surface waves, which have larger

Figure 96 *The San Francisco peninsula. The San Andreas Fault runs vertically through the center of the photograph.* (Photo by R. E. Wallace, courtesy USGS)

amplitudes and lower frequencies than body waves. Generally, the severity of the ground shaking increases with earthquake magnitude and decreases with distance from the epicenter. Body and surface waves cause buildings to vibrate in a complex manner. If part of a building moves in one direction while another part moves in a different direction, the damage is often catastrophic.

The building site also affects the amount of movement. Usually, structures built on bedrock (hard subsurface rock) are less severely damaged than those built on less consolidated, easily deformed material such as natural and artificial fills. Soft sediments generally absorb the high-frequency vibrations and amplify those of low frequency. Furthermore, the longer the period the ground is in motion, the more severe the earthquake damage. Aftershocks, which are subsequent minor or moderate earthquakes caused by readjust-

TABLE 8 MAJOR VOLCANIC DISASTERS OF THE 20TH CENTURY

Date	Volcano/Area	Fatalities	Remarks
1902	La Soufrière	15,000	
	Pelée (Martinique)	28,000	
	Santa María (Guatemala)	6,000	Deadliest volcano outbreak of 20th century, resulting in a total death toll of 35,000
1919	Keluit (Indonesia)	5,500	Deaths resulted from volcanic mudflows called lahars
1977	Nyiragongo (Zaire)	70	
1980	St. Helens (United States)	62	Worst volcanic disaster in the nation's recorded history
1982	Galunggung (Indonesia)	27	Relatively few deaths, but huge loss of property and suffering
1983	El Chichón (Mexico)	2,000	Worst volcanic disaster in the nation's recorded history
1985	Nevado del Ruiz (Colombia)	22,000	Worst volcanic disaster in the nation's recorded history
1986	Lake Nios (Cameroon)	20,000	
1991	Unzen (Japan)	37	Ongoing eruptions have forced 3,000 people form their homes
1991	Pinatubo (Philippines)	700	Largest eruption in nation's recorded history; remarkably few fatalities despite destruction
1993	Mayon (Philippines)	75	Eruptions produced deadly pyroclastic flows, 60,000 evacuated

ments in the rocks following the major shock, can be just as destructive and finish off what the main event began.

Certain structures are particularly susceptible to earthquakes because the frequencies of seismic waves often coincide with the structural resonant frequencies. All buildings can withstand large vertical stresses of one or more times Earth's gravitational acceleration, or g force, because they are built against the force of gravity. However, because the largest ground motions are usually in the horizontal direction, special precautions must be observed to ensure adequate resistance to large horizontally directed forces.

As the energy released by an earthquake travels along the surface, it vibrates the ground in a complex motion, moving up and down as well as from side to side. The destruction attributable to the vibrations depends on the intensity and duration of ground shaking, the nature of the material on which the structure rests, and the design of the structure itself. Steel-framed buildings survive better than rigid concrete ones, whereas flexible wood-framed buildings survive the best.

The size of the geographic area affected by ground shaking depends on the magnitude of the earthquake, which is proportional to the length and depth of the fault. Furthermore, the ground shaking close to a rupturing fault is more violent than previously thought. The seismic fling effect, which occurs only near the earthquake source, literally pulls the ground out from under structures, then yanks it quickly back again. This phenomenon has the potential to topple over some of the most seismic resistant buildings. For a major metropolitan area, a greater threat exists from a moderate earthquake on a minor fault directly beneath the city than from a major earthquake on a large fault some distance away.

Some types of ground transmit seismic energy more effectively than others. A comparison between western and eastern earthquakes in the United States shows that for a given intensity of ground shaking, damage extends over much wider areas in the East compared with the West (Fig. 97). Most earthquakes in the Appalachian region occur on a particular set of ancient Iapetan faults that formed more than 600 million years ago. At that time, the ancestral North American continent rifted away from the supercontinent Rodinia, forming the Iapetus Sea, a predecessor of the Atlantic. The northwest side of the Appalachian Mountains are unusually seismically active for unknown reasons.

Ground shaking affects a larger area in the East because seismic wave amplitudes decrease more slowly with increasing distance from the epicenter.

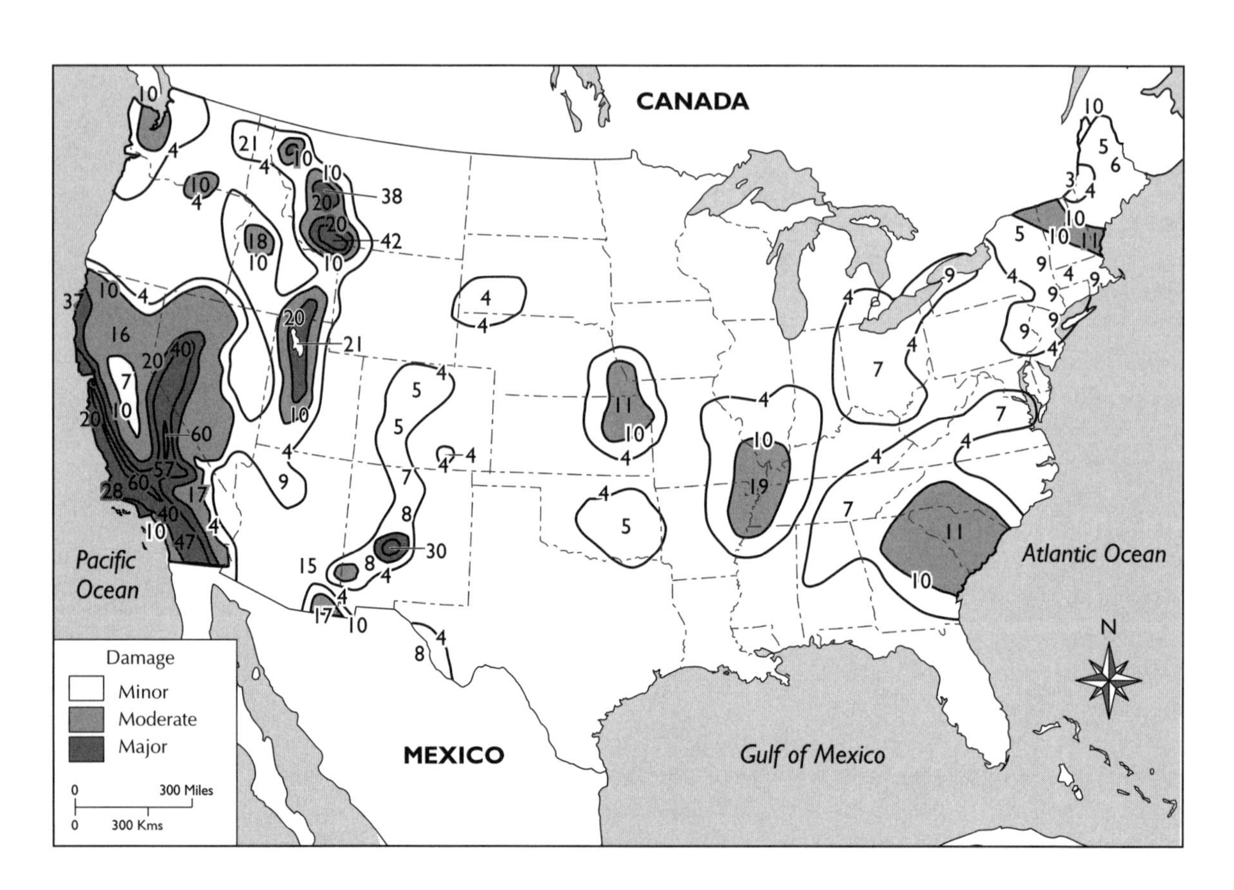

Figure 97 *Comparison between western and eastern earthquake area destruction.*

Therefore, earthquakes are felt over longer distances, indicating substantial differences in the crustal composition and structure of the two regions. The East contains older sedimentary structures. In contrast, the West comprises relatively young igneous and sedimentary rocks riddled with faults.

EARTHQUAKE DANGERS

The earthquake magnitude (Table 9) is proportional to the length and depth of the rupture created by slipping plates. Generally, the deeper and longer the fault, the larger the earthquake. An earthquake begins spontaneously. A break along a fault triggers a larger rupture, which in turn triggers an even larger one until finally the activity is significant enough to be damaging. All earthquakes begin slowly before releasing most of their energy in a burst. However, larger quakes take longer to get started. Other processes that affect earthquake magnitude include the frictional strength of the fault, the drop in stress across the fault, and the speed of the rupture as it traverses over the fault. A break along a fault can travel at speeds of up to 1 mile a second.

Earth's crust is constantly shifting, producing vertical and horizontal offsets on the surface. These movements are associated with large fracture zones in the crust. The greatest earthquakes have offsets of several tens of feet occurring in a matter of seconds. Most faults are associated with plate boundaries. Most earthquakes are generated in zones where huge plates collide or shear past each other. If the plates hang up in so-called stuck spots called asperites, the sudden release generates tremendous seismic energy. The interaction of plates causes rocks to strain and deform. If deformation takes place near the

TABLE 9 SUMMARY OF EARTHQUAKE PARAMETERS

Magnitude	Surface wave height (feet)	Length of fault affected (miles)	Diameter area quake is felt (miles)	Number of quakes per year
9	Largest earthquakes ever recorded—between 8 and 9			
8	300	500	750	1.5
7	30	25	500	15
6	3	5	280	150
5	0.3	1.9	190	1,500
4	0.03	0.8	100	15,000
3	0.003	0.3	20	150,000

surface, major earthquakes result. Earthquakes also occur during volcanic eruptions, but they are mild compared with those caused by faulting.

Thousands of earthquakes strike yearly. Fortunately, only a few are sufficiently powerful to be destructive (Table 10). During the 20th century, the world average was about 18 major earthquakes of magnitudes 7.0 or greater per year. For great earthquakes with magnitudes above 8.0, the century's average was 10 per decade. Damage does not depend on magnitude alone, which is proportional to the length and depth of the rupture, but is also influenced by the geology of the region.

Earthquakes occurring in strong rocks, such as those in continental interiors, are more destructive for equal measures of magnitudes than those occurring in the fractured rock at plate margins. This is why earthquakes in the eastern United States influence a wider area than those in the West. The August 31, 1886, Charleston, South Carolina, earthquake (Fig. 98), which killed 110 people locally, cracked walls in Chicago 750 miles away and was felt in Boston, Milwaukee, and New Orleans.

Generally, the longer the time since the last big shock on a major fault, the greater the earthquake hazard. The earthquake hazard rises as the interval since the last jolt on a major fault increases. This is known as the seismic gap hypothesis, which holds that the earthquake hazard along faults is low immediately following a large earthquake and increases with time. Because much time is needed for strain to build up again, only long-dormant faults are prime hazards. Although the theory applies only to large earthquakes, a moderate

TABLE 10 EARTHQUAKE MAGNITUDE SCALE AND EXPECTED INCIDENCE

Magnitude scale	Earthquake effects	Yearly average
< 2.0	Microearthquake—imperceptible	+600,000
2.0–2.9	Generally not felt but recorded	300,000
3.0–3.9	Felt by most people if nearby	50,000
4.0–4.9	Minor shock—damage slight and localized	6,000
5.0–5.9	Moderate shock—equivalent energy of atomic bomb	1,000
6.0–6.9	Large shock—possibly destructive in urban areas	120
7.0–7.9	Major earthquake—inflicts serious damage	14
8.0–8.9	Great earthquake—inflicts total destruction	1 a decade
9.0 and up	Largest earthquakes	1–2 a century

Figure 98 *Wreckage from the August 31, 1886, Charleston, South Carolina, earthquake.*

(Photo by J. K. Hillers, courtesy USGS)

quake could strike on the same fault without warning, making occurrences very unpredictable.

Other factors that influence the earthquake hazard are the size of the temblor and the geology of the region. Most faults appear to have a characteristic earthquake that recurs in a similar form. Some areas might experience similar earthquakes of 7.0 magnitude, whereas other regions might be prone to great earthquakes of 8 or 9 magnitude. However, larger quakes do not follow the same patterns set by smaller ones, making their prediction extremely tenuous. Earthquakes are likely to strike where they have occurred before. Once a zone becomes seismically active, earthquakes continue until, for unknown reasons, they cease. Then a relatively long interval passes before another great one strikes.

DAMAGE PREVENTION

The damage arising from a major earthquake is widespread, altering the landscape for thousands of square miles. The crust is constantly readjusting itself, resulting in vertical and horizontal displacements on the surface associated with fracture zones in the crust. Large earthquakes can produce offsets of several tens of feet in only a few tens of seconds. The rupturing faults can also communicate with other long-distant faults, triggering earthquakes up to thousands of miles away.

Besides the destruction of buildings and other structures, earthquakes alter the landscape by producing deep fissures and tall scarps and by causing massive landslides that scar the terrain. The greatest deformation occurs near thrust faults, where one block overrides another, especially on the hanging wall, which rises during an earthquake. Great thrust quakes break the crust diagonally and involve more surface area than other forms of faulting. Active faults, responsible for scarps, rifts, and mountain ranges, crisscross much of the land surface at plate boundaries on the edges of continents and in the continental interiors underlain by old rifts.

Ancient civilizations living in earthquake-prone regions have protected themselves from the ravages of quakes by constructing simple dwellings that withstand violent shaking. Today, however, as accommodations have become more sophisticated with complex construction materials earthquake damage has become a serious and expensive problem. In urban areas, the principal cause of earthquakes property damage is fires started by crossed electrical wires and ruptured natural-gas lines. The fires often burn out of control due to broken water mains and disrupted communications.

Most large urban centers are a combination of new construction blended in with old architecture, whose foundations have weakened with age. Buildings in earthquake-prone areas must withstand a major temblor during their lifetimes, especially structures vital to the community such as physical plants, hospitals, and schools. However, with space becoming a premium, designs and materials might not always strictly conform to earthquake-building codes due to rapid development in urban areas, where tall skyscrapers are erected.

The ability of buildings to withstand ground shaking depends on the type of ground being built on, the building design, the type of building materials, the quality of construction, the orientation with respect to the shock wave, and the nature of the earthquake shock. A short, sharp, high-frequency shock of only a few seconds duration is comparatively easy to design for. Two- to four-story buildings are most vulnerable to this type of shock, whereas taller buildings generally escape major destruction. However, a longer, lower-frequency shock of up to a minute or more in duration is more difficult to design for. Multistory buildings tend to tumble, while shorter buildings are practically unscathed.

Scientists from many disciplines have advanced theories on earthquake construction and design. Their ideas were put to the test during the 1971 San Fernando earthquake of magnitude 6.6 and 60-second duration. Single-story houses performed better than multistory buildings. Whether floors were constructed of wood or concrete slabs made little difference. Newer houses survived on the whole better than older ones, and flimsily built mobile homes were often heavily damaged. Even schools failed to stand up to the tremors. Freeways were twisted and contorted, and newly constructed overpasses collapsed (Fig. 99).

Prior to the earthquake, building codes were established on the premise of a probable maximum acceleration of 10 percent of gravity for a moderate earthquake. Nevertheless, when accelerographs in the San Fernando Valley indicated the exposure of some buildings to accelerations equal to or greater than the force of gravity, the codes were found grossly inadequate. If a sizable earthquake had struck the area, buildings would have been subjected to forces five to 10 times stronger than anticipated in the original specifications.

Furthermore, the longer the duration of ground shaking, the greater the possibility that tall buildings will resonate and start to swing, devastating a building designed to withstand only a short-duration earthquake. Geologists

Figure 99 *Collapsed freeway overpass from the February 9, 1971, San Fernando, California, earthquake.*

(Photo courtesy USGS)

have located most of the active faults. This gives engineers information on the direction seismic waves are likely to originate. This allows engineers to orient structures with the long axis aligned parallel to the expected ground motion to prevent serious damage.

Many areas such as the Pacific Northwest, which lies along a subduction zone and has been devastated by earthquakes in the prehistoric past, were not built to survive severe ground shaking and are totally unprepared for a major quake. About 1,000 years ago, a huge earthquake struck the Seattle, Washington, area. The ground shook with such fury that avalanches and landslides tumbled from the Olympic Mountains and buried areas that are now densely populated. The earthquake also triggered a great tsunami that washed the shores of Puget Sound.

The type of building construction also determines how well a structure survives an earthquake. The January 26, 2001, earthquake of 7.9 magnitude in the heavily populated region of Gujarat in western India killed some 30,000 people, mostly when poorly constructed buildings collapsed upon them. Lightweight, steel-framed buildings with strength combined with flexibility would be expected to suffer little damage. Reinforced concrete buildings with few door or window openings that tend to weaken the structure also have a good chance of survival. Unreinforced hollow concrete block structures and older brick buildings are generally heavily damaged (Fig. 100).

Modern buildings propped up on concrete pillars with inadequate cross-bracing to make room for parking lots below create a structure doomed to fail

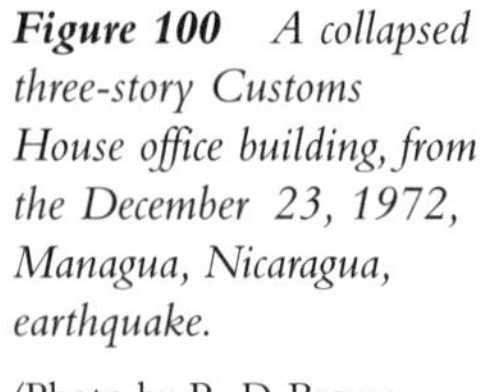

Figure 100 *A collapsed three-story Customs House office building, from the December 23, 1972, Managua, Nicaragua, earthquake.*

(Photo by R. D. Brown, courtesy USGS)

even in the slightest tremor. Overweight roofs tend to make buildings top-heavy. If balconies and parapets break loose, they endanger bystanders in the streets below. Consequently, people are urged to remain inside well-constructed buildings during an earthquake instead of rushing out into the streets.

Earthquake-proof construction is a new technology, but certain basic rules have been known for some time. Unreinforced concrete or masonry buildings are among the most vulnerable, whereas wood-framed, single-story houses are among the safest. Two schools of thought concern multistory, steel-framed buildings. American engineers believe buildings should be designed as flexible as possible, whereas Japanese engineers prefer buildings to be as rigid as possible. While skyscrapers with flexible steel skeletons might remain standing through an earthquake, they have the added hazard of wrenching violently out of plumb. Some buildings sway so fiercely during an earthquake that they seriously injure occupants as furniture and other objects are hurled against walls.

Earthquake engineering deals with the efficient and economic design of structures subjected to ground shaking. Research and development in areas such as earthquake engineering, risk analysis, land use regulation, stricter building codes, and disaster preparedness reduce casualties and lower economic losses. Also included is the renovation of existing structures to improve performance in an earthquake. Houses, commercial buildings, schools, hospitals, dams, bridges, and power plants are closely examined for weaknesses, which could be fatal to the community when an earthquake strikes.

Even if structures withstand the initial ground shaking, the added hazard of foundation failure could cause buildings to topple over. Severe ground shaking can cause certain types of soils to settle or liquefy, thereby losing their structural support. Countermeasures can prevent ground failure, including the drainage of the affected area or the institution of land use regulations that limit construction on vulnerable soils, along active faults, in potential landslide areas, or in coastal zones subject to destructive tsunamis (Fig. 101).

Risk assessment is an important part of earthquake research. It determines to what extent are the added costs of safer construction offset by the potential of saving life, property, and productivity in an area given a probability of strong earthquakes. Experience has shown that people remain in an area for various reasons no matter how hazardous. Cities are often established for their climate, economic importance, strategic defense, and recreational facilities. When natural or man-made disasters destroy cities, they are rebuilt on the original site for the same reasons they were built there in the first place. All these factors, including the hazards of fire and flooding, are combined in an overall assessment of the risk on which decisions are based for land use and construction practices in earthquake-prone regions.

The cost of earthquakes in terms of death and destruction has risen dramatically in recent years. Although the January 17, 1995, Northridge, California, earthquake killed only 63 people, it caused upward of $13 billion in damages (Fig. 102). Steel-framed buildings, which are considered more pliant than ones built with steel-reinforced concrete and therefore more resistant to powerful earthquakes, suffered more harm in the Northridge quake than had been thought.

A disaster similar to the 1906 San Francisco, California, earthquake today could cost from $115 billion to $135 billion and kill 2,000 to 6,000 people. A magnitude 7 temblor in the Los Angeles basin could cause $125 billion to $145 billion in damages and kill 2,000 to 5,000 people. A repeat of the 1923 Tokyo, Japan, earthquake could cause 40,000 to 60,000 deaths and wreak damages amounting to between $800 billion and $1.2 trillion, which would devastate the Japanese economy. Therefore, earthquakes are a well-recognized hazard to many of the world's major cities.

EARTHQUAKE PREDICTION

Earthquake prediction relies on the development of integrated systems of measurement, rapid automated data collection and analysis, and the application of

Figure 101 *Houses and debris carried by a tsunami into the lagoon area at the north end of Seward, Alaska, from the March 27, 1964, Alaskan earthquake.*

(Photo by courtesy Alaska Earthquake USGS)

Figure 102 *Buildings damaged in the Marina district, San Francisco from the October 17, 1989, Loma Prieta, California, earthquake.*
(Photo by G. Plafker, courtesy USGS)

this information to the complexities of crustal deformation. Scientific investigations indicate that strong earthquakes might be predicted years in advance. Methods for making short-term predictions of weeks or even days could also be developed. Despite some successful predictions, however, most seismologists have given up predicting earthquakes after failing to find reliable warning signs preceding quakes.

The long-term evidence might suggest a major earthquake is imminent, yet the short-term geophysical parameters remain essentially unexplained and confusing. They could be precursors of an earthquake or simply normal fluctuations in the crust. The long-term warning allows time to devise a remedial policy that would greatly reduce casualties and lower property damage. The long-range prediction could encourage the strengthening of existing structures in the threatened area as well as motivate authorities to revise and enforce building and land use codes.

The short-term predictions could mobilize disaster relief operations and set in motion procedures for evacuating weak or flammable structures and other dangerous areas. Shutting down hazardous facilities, such as nuclear power plants, petroleum refineries, and natural gas pumping stations, and evacuating areas below dams and low-lying coastal regions subject to floods, landslides, or tsunamis would result from a short-term prediction.

Evacuation in most cases might be impractical, even if a warning is given. It might even result in more deaths from traffic jams than the earthquake itself.

The only other practical alternative is to construct earthquake-proof structures that can withstand minor earthquakes with little or no damage, survive moderate earthquakes with only minor damage, and provide a degree of safety in large earthquakes, even though the building is heavily damaged.

An earthquake prediction with the same accuracy as a weather forecast is desirable because large earthquakes can potentially cause more damage to urban areas than hurricanes, tornadoes, or floods. Many earthquake precursors need to be found. Each would be based on a different physical parameter. This would enhance confidence in earthquake prediction when based on several independent lines of evidence.

The plate tectonics model, which explains the occurrence of major earthquakes along the boundaries of crustal plates, combined with earthquake statistics make prediction possible by identifying particularly dangerous areas. The information helps establish a historical record for an earthquake-prone region and estimate the relative danger. When averaged over a sufficiently long period, the sum of the slippage along faults should equal the total displacement between the two plates.

Geologic fingerprints left by previous earthquakes in the strata around the San Andreas Fault (Figs. 103a, b) date as far back as 20 million years ago. The fault has roots that extend 15 miles through the crust, reaching the top of the mantle. Past earthquakes offset old stream channels on one side of the fault with respect to the other side. By measuring the displacement and dating the strata, a relative magnitude and date can be obtained. The method has uncovered a 1,400-year record, during which 12 large earthquakes have occurred. The interval between earthquakes is from 50 to 300 years, with an average interval of roughly 150 years.

Nearly 140 years have elapsed since the last great earthquake on the southern San Andreas Fault. Moreover, the rate of strain accumulating along the fault is roughly 1.5 inches per year, making the total strain about 15 feet, which could generate an earthquake of magnitude 7.5 or more. An estimate of the probability of an earthquake of magnitude 8.0 or greater occurring along the southern San Andreas Fault is between 2 and 5 percent per year, rising to perhaps 50 percent in 20 to 30 years.

Seismically active regions produce more small-to-moderate earthquakes than large ones. Large earthquakes in California have occurred in the past on average every 150 years or so. Moderate quakes occur about every 22 years, and numerous small tremors occur yearly. Periods of calm before a strong shock are frequently observed, with seismic activity dropping to a minimum and then increasing dramatically just before the main shock.

Large earthquakes can apparently provide warning times of roughly 10 years or more. The magnitude of the predicted earthquake depends on the duration of the precursor anomalies. For instance, an event with a magnitude

Figure 103a *The San Andreas Fault in California.*

(Photo by R. E. Wallace, courtesy NASA)

of 5.0 has an anomaly lasting about four months. A major earthquake of 7.0 with about 1,000 times more energy could be preceded by an anomaly beginning several years before the event.

The discovery that the size of an earthquake as well as its location and timing is predictable should be important for land use planning. Furthermore, the larger the magnitude of the forthcoming quake, the longer the lead time available for planning to combat its effects. Unfortunately, nature does not always act accordingly. Although time might be available for preparations, an earthquake could strike when least expected.

Thus far, only a few earthquakes have been successfully forecasted. Many others have been "predicted" after the fact by reviewing the data and finding the precursory signals. How many formal predictions based on the methods described have failed is undeterminable because studies have been conducted only over the last few decades and the science is not yet perfected. Therefore,

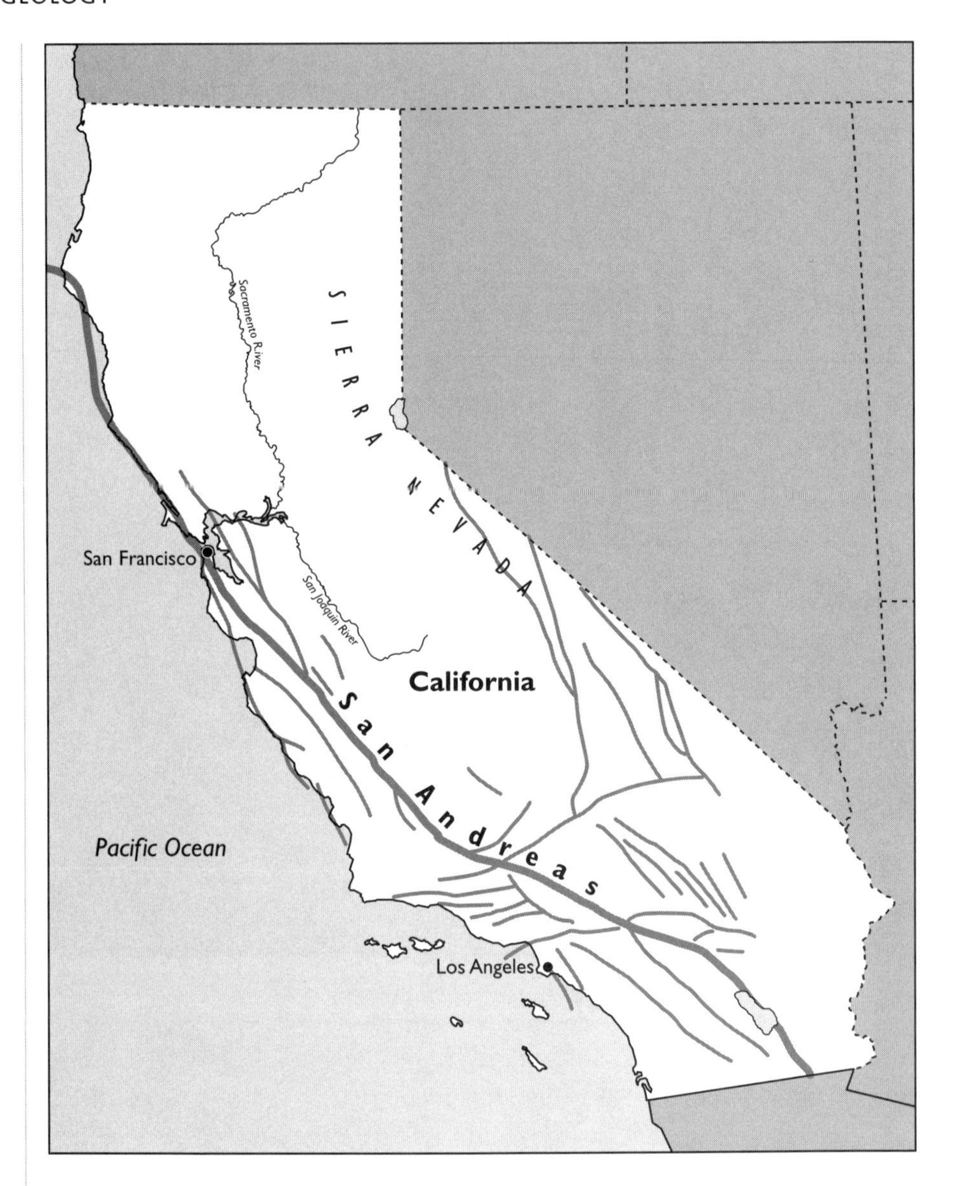

Figure 103b *The San Andreas and associated faults.*

basing a worldwide earthquake watch using the rudimentary knowledge scientists have gained thus far might be premature.

Although major earthquake belts extend for tens of thousands of miles, only a small fraction is adequately instrumented to test prediction methods. With the pooling of data gathered in various parts of the world, tests of the validity of prediction methods are constantly being evaluated. The purpose is to interpret the signals of strain accumulating in the crust for a better understanding of the earthquake processes.

Many potentially important methods are tested. These include arrays of wells that monitor water level and radon gas content, networks of resistivity sensors, magnetometers and gravity meters (Fig. 104 and 105), sea-level gauges, and advance surveying techniques. Therefore, physical parameters of the rocks in faults, such as electrical resistivity, seismic wave velocity, and the amount of deformation, can be monitored for earthquake prediction, just as temperature, pressure, and wind direction are monitored to forecast the weather.

The leading agency for earthquake prediction in the United States is the Geological Survey. In central California, investigators have installed a network of stations equipped with seismometers and tiltmeters along the San Andreas and associated faults. Magnetic and electrical observations are also conducted. In addition, the instrumentation includes several networks of laser distance-ranging devices.

Vertical strain in the crust corresponds to changes in the elevation of the land. This is measured directly by standard survey methods and indirectly by determining the local strength of Earth's gravitational field. Geodetic measurements from Global Positioning Satellites offer measurements of horizontal movement of crustal blocks adequate for earthquake prediction. A

Figure 104 *A magnetometer survey near Silver Spring, Montgomery County, Maryland.*

(Photo by E. F. Patterson, courtesy USGS)

Figure 105 *A gravimeter survey in southern California in 1978.*

(Photo courtesy USGS)

considerable effort has gone into developing reliable, inexpensive instruments that measure strain continuously at a single point on Earth.

A system involving a network of sensors that detect seismic waves before they spread from the epicenter of an earthquake can provide a few seconds of advance warning to outlying areas. A central computer would process data transmitted by seismic sensors to determine the size and location of the quake and send out information to areas in the path of damaging vibrations. The early warning would activate automated systems to shut off electrical power and natural gas substations, reducing the risk of damaging fires. The system would also provide information to help emergency officials quickly locate the sites hardest hit by earthquakes, mobilizing rescue crews in a timely manner to save lives and property.

VOLCANIC ACTIVITY

Most of the world's active volcanoes concentrate in a few narrow belts (Fig. 106). A nearly continuous zone of volcanism called the Ring of Fire surrounds the Pacific Basin, wherein lies three-quarters of the world's active volcanoes. The region coincides with the circum-Pacific belt because the same tectonic processes that produce earthquakes also create volcanoes. The volcanism is associated with a band of subduction zones around the Pacific rim. As oceanic crust subducts into the mantle, it melts to provide molten magma for volcanoes fringing the deep-sea trenches. Consequently, most of the active volcanoes in the world are in the Pacific Ocean, with nearly half residing in the western Pacific region alone.

Subduction-zone volcanoes, such as those in Indonesia (Fig. 107) and the western Pacific, are among the most explosive in the world. Their violent behavior results from large amounts of volatiles in their magmas, consisting of water and gases. As the magma rises toward the surface, the pressure drops. Volatiles quickly escape, shooting out of the volcano as though propelled by a gigantic canon. Rift volcanoes form when the divergence of lithospheric plates exposes the upper mantle to the surface at midocean ridges and continental rifts such as the East African Rift and Iceland.

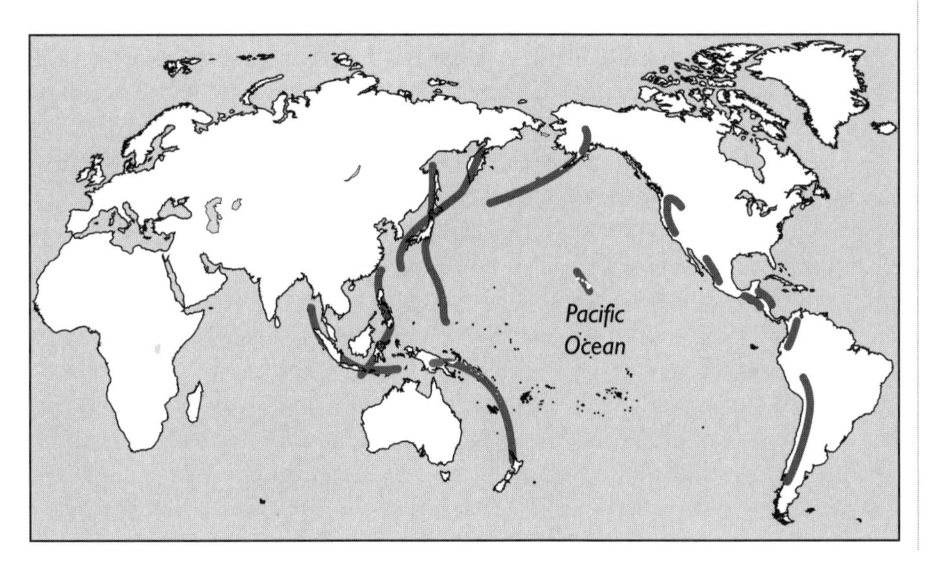

Figure 106 *Major belts of active volcanoes.*

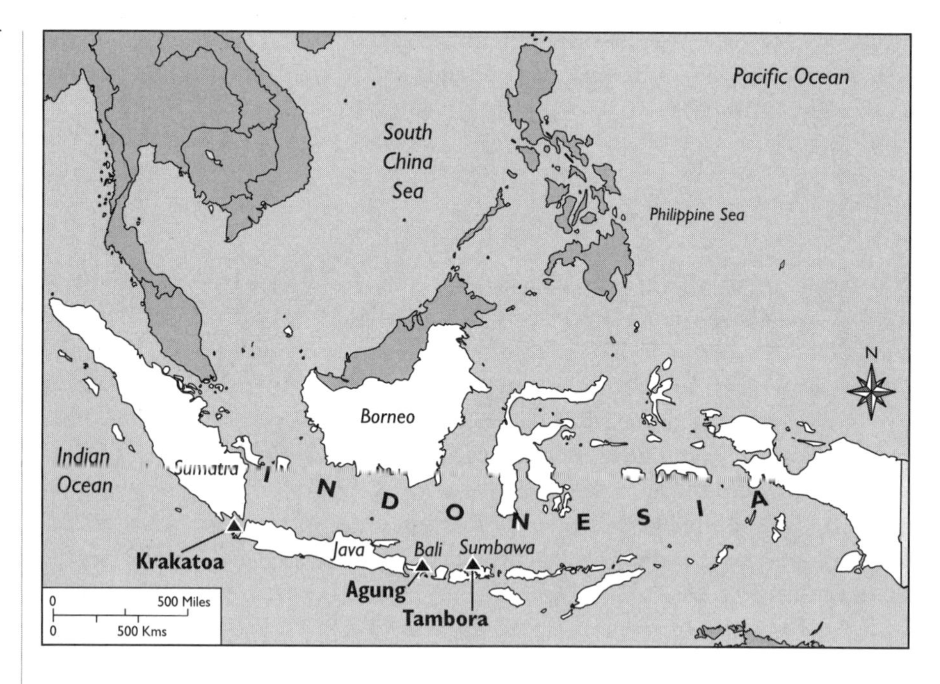

Figure 107 *Location of the great Indonesian volcanoes.*

More than 100 small regions of isolated volcanic activity scattered around the world are known as hot spots. They generally lie in the interiors of plates far from plate margins, where most world's active volcanoes occur. Hot-spot volcanoes derive their magma from deep within the mantle, possibly from the very top of the core. The magma rises in giant mantle plumes that provide a steady flow of molten rock to magma chambers.

The composition of the magma determines its viscosity and whether it erupts mildly or explosively. If the magma is highly fluid and contains little dissolved gas when it reaches the surface, it produces basaltic lava. The eruption is usually quite mild. However, when encroaching upon towns and villages, basalt flows can be quite destructive, as the citizens of Parícutin, Mexico, discovered on June 10, 1943 (Fig. 108). The two types of lava from this type of eruption are aa, or blocky lava, and pahoehoe, or ropy lava, which are Hawaiian names and typical of Hawaiian volcanoes. However, if the magma rising toward the surface contains a large quantity of dissolved gases, it erupts in a highly explosive manner that can be quite destructive.

Volcanic islands began as undersea volcanoes. Explosive eruptions associated with the creation or the destruction of a volcanic island also set up large tsunamis. Volcanic eruptions that develop tsunamis are responsible for about one-quarter of all deaths caused by tsunamis. The powerful waves transmit the volcano's energy to areas outside the reach of the volcano itself. Large pyroclastic flows entering the sea or landslides triggered by volcanic eruptions also produce destructive tsunamis.

In 1792, during the earthquake following the eruption of Unzen, Japan, one side of the volcano collapsed into the bay. This created an enormous tsunami up to 180 feet high. It washed coastal cities out to sea. Up to 15,000 people vanished without a trace. The volcanic island of Krakatoa, between Java and Sumatra, was nearly totally decimated on August 27, 1883, by a series of tremendous explosions that collapsed most of the island. In the nearby coastal areas, the eruption produced towering tsunamis more than 100 feet high that swept 36,000 people to their deaths.

Alaska's Mount St. Augustine (Fig. 109) has often collapsed and fallen into the sea, generating huge tsunamis. Massive landslides have ripped out the flanks of the volcano 10 or more times during the past 2,000 years. The last slide occurred during the October 6, 1883, eruption. Debris on the flanks of the volcano crashed into the Cook Inlet, sending a 30-foot tsunami to Port Graham 54 miles away that destroyed boats and flooded houses. Subsequent eruptions have filled the gap left by the last landslide, destabilizing the volcano once again, making another collapse imminent. If a landslide did occur, it would barrel down the north side of the volcano and plunge into the sea, unleashing a tsunami in the direction of cities and oil platforms residing in the inlet.

In volcanic mountainous regions, seismic activity and uplift accompanying eruption cause landslides in thick deposits of pyroclastic material on a

Figure 108 *Lava from the June 10, 1943, eruption of Parícutin Volcano buried a nearby village, Michoacán, Mexico.*

(Photo courtesy USGS)

volcano's flanks. The distribution of landslides in volcanic terrain is determined by the seismic intensity, topographic amplification of the ground motion, the rock type, slope steepness, and fractures and other weaknesses in the rock. Heavy sustained rainfall over a wide area also triggers landslides and mudflows.

Mudflows associated with volcanic eruptions are called lahars, from the Indonesian word for mudflow because of their common occurrence in the region. Lahars are masses of water-saturated rock debris that descend the steep slopes of volcanoes similar to the flowage of wet concrete. The debris originates from loose, unstable rock deposited onto the volcano's flanks by explosive eruptions. The water comes from rain, melting snow, a crater lake, or a reservoir next to the volcano.

A tragic example of a lahar resulted from the 1919 eruption of Kelut Volcano on Java. It blew out the crater lake at its summit and created a large mudflow that killed 5,000 people. Lahars can also be initiated by pyroclastic or lava flows moving across a glacier and rapidly melting it. The best example in the United States was the May 18, 1980, eruption of Mount St. Helens, whose melted snowpack created many destructive mudflows (Fig. 110).

The lahar's speed mostly depends on the fluidity of the mudflow and the slope of the terrain. Lahars travel swiftly down valley floors. They often exceed 20 miles per hour while covering a distance of up to 50 miles or more. Lava flows extending onto glacial ice or snowfields produce floods as well as lahars. Flood-hazard zones for the volcanoes in the western Cascade Range extend long distances down some valleys and reach as far as the

Figure 109 *Mount St. Augustine, Kamishak district, Cook Inlet, Alaska.*

(Photo by C. W. Purington, courtesy USGS)

Figure 110 *A vehicle wrapped around a tree due to the force of a mudflow on the North Fork of Toutle River during the Mount St. Helens eruption on July 11, 1980, Cowlitz County, Washington.*

(Photo courtesy USGS)

Pacific Ocean. The vast carrying power of lahars can easily sweep away people and buildings.

The November 13, 1985, eruption of Nevado del Ruiz in Colombia, South America, melted the volcano's ice cap. It sent floods and mudflows cascading down the mountainside into the nearby Lagunilla and Chinchina River Valleys. The mudflow had a consistency of mixed concrete and carried off everything in its path. It buried almost all the city of Armero 30 miles away and badly damaged 13 smaller towns, leaving in its wake the deaths of some 25,000 people.

DANGEROUS VOLCANOES

Since the end of the Ice Age 12,000 years ago, some 1,300 volcanoes have erupted. During the past 400 years, over 500 volcanoes have killed more than 200,000 people and damaged billions of dollars worth of property. Since the year 1700, two dozen volcanoes have earned special recognition by killing over 1,000 people each (Fig. 111). Each of the past three centuries has shown a doubling of fatal volcanic eruptions. Recent decades have averaged about three deadly eruptions every year. Much of this increase is due to population growth and not necessarily from a greater number of deadly eruptions.

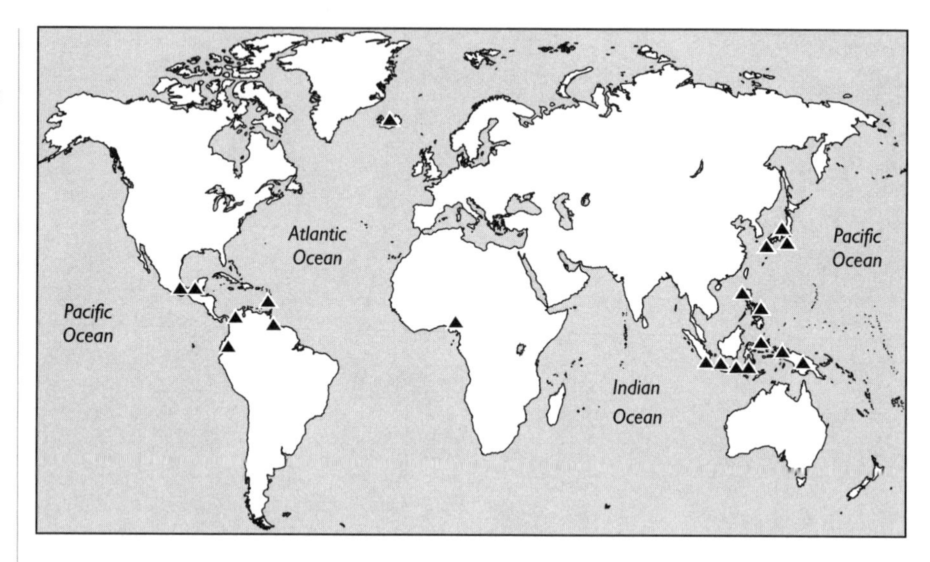

Figure 111 *The most dangerous volcanoes responsible for killing more than 1,000 people each since 1700.*

Pyroclastic flows, involving hot clouds of ash that sweep down the flanks of a volcano at hurricane speeds, are the predominant killer, claiming almost 80,000 victims. Volcano-triggered tsunamis, which threaten those living on coasts even hundreds of miles away from an eruption, have drowned about 55,000 people. Also among the major killers are mudflows and tephra, including ash, rocks, and other material thrown skyward by volcanoes. As the world's population grows out of control and people invade the domain of active volcanoes, death tolls will continue to rise. Developing countries, deprived of the technology to provide advance warning, will suffer the highest number of deaths from volcanoes.

Several volcanoes in the American West are poised for eruption at anytime. Many could be awaking from long slumbers. Even 50,000 years of inactivity is not enough time to silence the rumblings of volcanoes, many of which have awakened even after more than 1 million years of sleep. Predicting future eruptions requires a determination of a volcano's past behavior by studying its rocks. The volcano can then be grouped along with others in a descending order of hazard.

Some 35 volcanoes in the United States, mostly in the Cascade Range (Fig. 112), are likely to erupt sometime in the future. The largest known eruption occurred in the Cascades about 7,700 years ago when the 12,000-foot-tall Mount Mazama exploded. It left behind a huge, gaping hole filled with water to create Crater Lake, the deepest body of water in North America. The most hazardous volcanoes (Table 11) are those that have erupted on average every 200 years, that have erupted in the past 300 years,

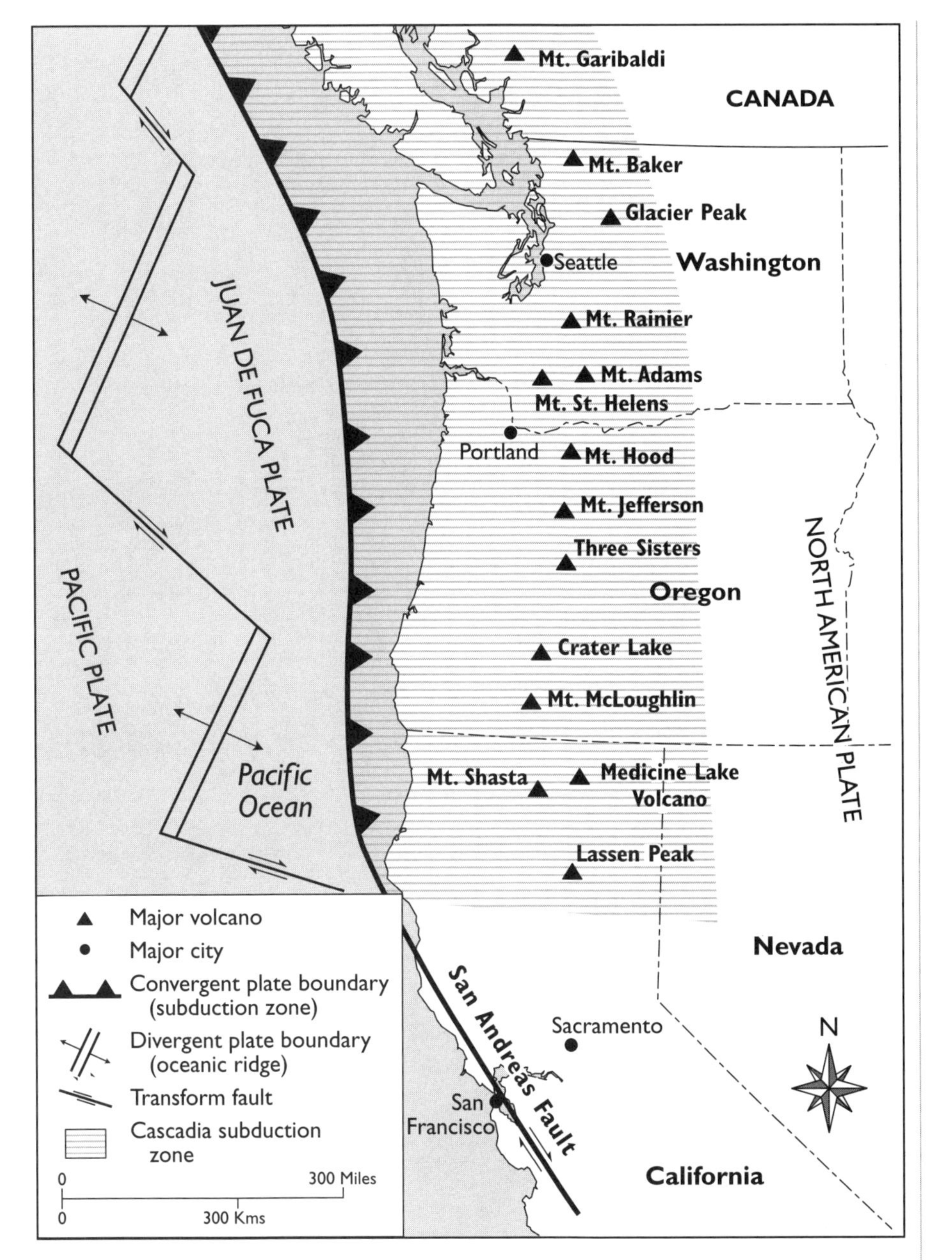

Figure 112 *Active volcanoes in the Cascade Range of the Pacific Northwest.*

or both. They include in order of rank Mount St. Helens, the Mono-Inyo Craters, Lassen Peak, Mount Shasta, Mount Rainier, Mount Baker, and Mount Hood.

TABLE 11 LOCATION OF THE MOST HAZARDOUS VOLCANOES IN THE UNITED STATES

Alaska

1. Augustine Volcano
2. Aleutian Volcanoes
3. Iliama Volcano
4. Katmai Volcano
5. Mount Edgecumbe
6. Mount Spurr
7. Mount Wrangell
8. Redoubt Volcano

Arizona

9. San Francisco Peak

California

10. Clear Lake Volcano
11. Coso Volcanoes
12. Lassen Peak
13. Long Valley Caldera
14. Mono-Inyo Craters
15. Mount Shasta

Hawaii

17. Haleakala
18. Haulalal
19. Kilauea
20. Mauna Loa

New Mexico

21. Socorro

Oregon

22. Crater Lake
23. Newberry Volcano
24. Mount Hood
25. Mount Jefferson
26. Mount McLoughlin
27. Three Sisters

Washington

28. Glacier Peak
29. Mount Baker
30. Mount Adams
31. Mount Rainier
32. Mount St. Helens

Wyoming

33. Yellowstone Caldera

Mount St. Helens was an old dormant stratovolcano with an almost perfectly symmetrical cone. It lies in the Cascade Range, stretching from northern California to southern British Columbia, along with 15 other major active volcanoes. It has erupted at least 20 times in the past 4,500 years. The last eruption of any consequence was in 1857. The most recent violent explosion occurred on May 18, 1980 (Fig. 113). It was the largest volcanic eruption in the continental United States in several centuries. The convulsion blew off the top one-third of the mountain and lofted a cubic mile of weather-altering debris into the atmosphere.

When Mount St. Helens erupted, entire slopes of the volcano collapsed. The volcano produced perhaps the largest landslides in recorded history. It created a massive mudflow and flood that reached all the way to the Pacific Ocean. The destruction was beyond imagination with more than 200 square miles totally devastated by the volcano. Total damage was estimated at nearly $3 billion, including enough timber to build 80,000 houses when the nearby forest was flattened by the blast.

Historical records indicate that prior to the 1980 eruption of Mount St. Helens, only two other volcanoes have erupted in the Cascade Range during the last century. A minor amount of ash erupted at Mount Hood, Oregon, in 1906. Several spectacular eruptions of Lassen Peak, California, occurred between 1914 and 1917 (Fig. 114). During the 19th century, Mounts Baker, Rainier, St. Helens, and Hood erupted ash or lava between 1832 and 1880. Periods between eruptions were 10 to 30 years for each volcano, with perhaps as many as three volcanoes erupting in the same year.

The next most hazardous are volcanoes that erupt less frequently than every 1,000 years and last erupted more than 1,000 years ago. They include Three Sisters, Newberry Volcano, Medicine Lake Volcano, Crater Lake Volcano, Glacier Peak, Mount Adams, Mount Jefferson, and Mount McLoughlin. The third most hazardous are volcanoes that last erupted more than 10,000 years ago but still overlie large magma chambers. They include Yellowstone Caldera, Long Valley Caldera, Clear Lake Volcanoes, Coso Volcanoes, San Francisco Peak, and Socorro, New Mexico.

The Long Valley Caldera (Fig. 115) east of Yosemite National Park, California, was created by a cataclysmic eruption 700,000 years ago. This resulted in a 20-mile-long, 10-mile-wide, and 2-mile-deep depression. Magma again appears to be moving into the resurgent caldera from a depth of several miles beneath the surface. The increased volcanic and seismic

Figure 113 *Eruption of Mount St. Helens on May 18, 1980.*

(Photo courtesy USGS)

Figure 114 *The 1914 eruption of Lassen Peak, Shasta County, California.*

(Photo by B. F. Loomis, courtesy USGS)

activity is indicated by a rise in the center of the caldera's floor of 1 foot or more since 1980.

Several medium-sized earthquakes of magnitude 6 or less have struck the region over the same period. These quakes appear to signal that magma is pressing toward the surface and that the caldera is poised for its first eruption in 40,000 years. Mammoth Mountain, a young volcano within the caldera, has experienced an extended period of activity and might be ready for an eruption. When this happens, thick basalt flows could flood large portions of neighboring Nevada.

A map of geologically recent eruptions shows 75 centers of volcanic activity arrayed in broad bands. They extend from the Cascade Range in northern California, Oregon, and Washington eastward through Idaho to Yellowstone and along the border between California and Nevada. Another band extends from southeast Utah through Arizona and New Mexico. All centers of activity have the potential for future eruptions. New centers of activity might form within these bands at any time.

PREDICTING ERUPTIONS

The same technology used for earthquake prediction can also be applied to predict volcanic eruptions. One method of predicting future eruptions is to

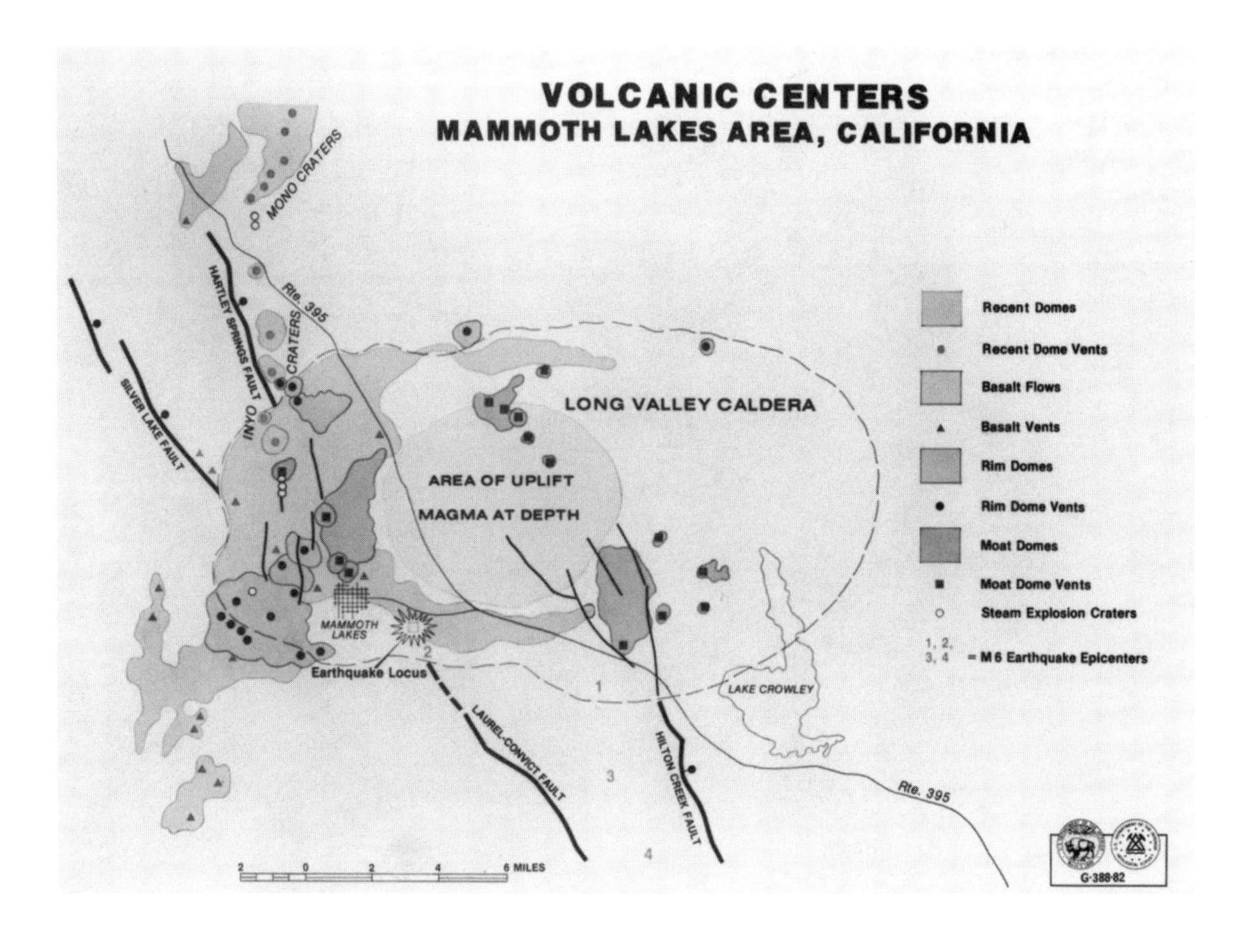

Figure 115 *Long Valley Caldera, showing volcanic centers of the Mammoth Lakes area, California.*
(Photo courtesy USGS)

determine a volcano's past behavior by studying its rocks. Foreseeing when a volcano will erupt is no more accurate than predicting earthquakes, however. As with earthquakes, some warning can be provided by seismographs that measure the intensity of seismic activity near the volcano.

In many volcanic regions, these preliminary quakes are accompanied by deep rumblings and by landslides cascading down the crater walls. The tilting of the ground around the volcano might prove to be the most reliable clue to impending activity. The underground movement of magma causes rapidly changing tilting of the land surface, which can be measured by tiltmeters (Fig. 116) placed strategically around the volcano.

The volcano might bulge along one of its flanks or begin to heave up the crater floor. The heat generated within the volcano might melt the ice and snow on its summit. As the magma rises in volcanic vents, it distorts Earth's magnetic field around it, which can be measured by airborne magnetometers. Along with the magnetic changes, alterations in electric currents in the ground can be measured by sensitive resistivity meters. Other indicators of

Figure 116 *A semiportable tiltmeter built at the Hawaiian Volcano Observatory and used for volcano investigations locally and in the Aleutian Islands.*

(Photo by R. H. Finch, courtesy USGS)

Figure 117 *Seawater is being sprayed directly onto a lava flow in the outer harbor of Vestmannaeyjar, Iceland, from the May 4, 1973, eruption on Heimaey to arrest it from infilling the harbor entrance.*

(Photo courtesy USGS)

impending eruption include a sudden rise in the temperature of nearby hot springs, gas vents, and near-surface rocks.

With sufficient warning, the damage caused by a volcanic eruption might be reduced. Channels can be dug and levees built to divert the flow of lava away from inhabited areas. The lava stream can be doused with water to cool it, causing the flow to slow and solidify. Seawater was successfully sprayed onto a lava flow in the outer harbor of Vestmannaeyjar, Iceland, during the eruption of Heimaey Volcano in May 1973 in an attempt to slow its advance into the sea (Fig. 117). The Air Force has even attempted to bomb lava flows from Mauna Loa in Hawaii with some success.

Dams were built in Java to divert volcanic mudflows away from cities and agricultural lands. Some villages built artificial hillocks, which served as islands of refuge from volcanic mudflows. When Mount Etna produced a spectacular eruption in April 1992, lava flows threatened the Sicilian town of Zafferana Etnea, whose townspeople battled desperately against the fury of the volcano to turn to lava away from their homes. People who make active volcanoes their neighbors learn to live with their unruly behavior and treat them accordingly as though they were just a normal part of their lives.

After discussing the effects earthquakes and volcanoes have on civilization, the next chapter concentrates on the destructive forces of erosion, slides, and ground failures.

7

LOSING GROUND

EROSION AND SLIDES

This chapter examines erosional processes, earth movements, and ground failures. Human activity has produced geologic effects of an extensive and alarming character. Over geologic time, powerful erosional forces have obliterated many of Earth's features, including ancient cities long abandoned because of the abuse of the land. History shows that soil erosion has been a continuing problem down through the ages. Erosion destroys the land and retards the progress of civilization. Soil erosion continues to be the most critical limiting factor to further human population growth.

Landslides and related phenomena are often destructive, taking many lives and ravaging the land. Slides are rapid downslope movements of soil and rock materials triggered mainly by seismic activity and severe storms. Most slides are not as spectacular as other violent forms of nature. Nonetheless, they are more widespread, causing major economic losses and casualties in virtually every region of the world. Furthermore, they accompany other geologic hazards, including earthquakes, volcanic eruptions, and floods, contributing to their mayhem and destruction.

SOIL EROSION

Soil is one of the most endangered resources. Soil erosion (Fig. 118 and 119) removes from production millions of acres of once fertile cropland and pasture every year. Natural processes require thousands of years to generate a single inch of topsoil, an amount that is presently being lost in less than a decade. This situation is particularly distressing since the average soil depth worldwide is only 7 inches. World food production per capita will eventually fall off if the loss of topsoil continues. Therefore, short-term efforts to increase crop production to feed increasing populations are ultimately defeated if, in the long run, the topsoil erodes away.

The greatest limiting factor to continuing human population growth is soil erosion. This forces the world's farmers to feed more people on less topsoil. As much as a third of the global cropland is losing soil at a rate that undermines any long-term agricultural productivity. In other words, humans are "mining" the world's soils faster than nature is putting the earth back into the ground.

Before the arrival of agriculture around 10,000 years ago, natural soil erosion rates rarely exceeded 10 billion tons annually, slow enough for new soil to be generated in its place. However, present soil erosion rates are estimated at about 20 billion tons per year, equivalent to the loss some 15 million acres of arable land. Therefore, the world is losing soil twice as fast as it is being

Figure 118 *Sheet and rill erosion on this soybean field in Pittsylvania County, Virginia, causes soil losses of 80 to 100 tons per acre per year.*

(Photo by Tim McCabe, courtesy USDA Soil Conservation Service)

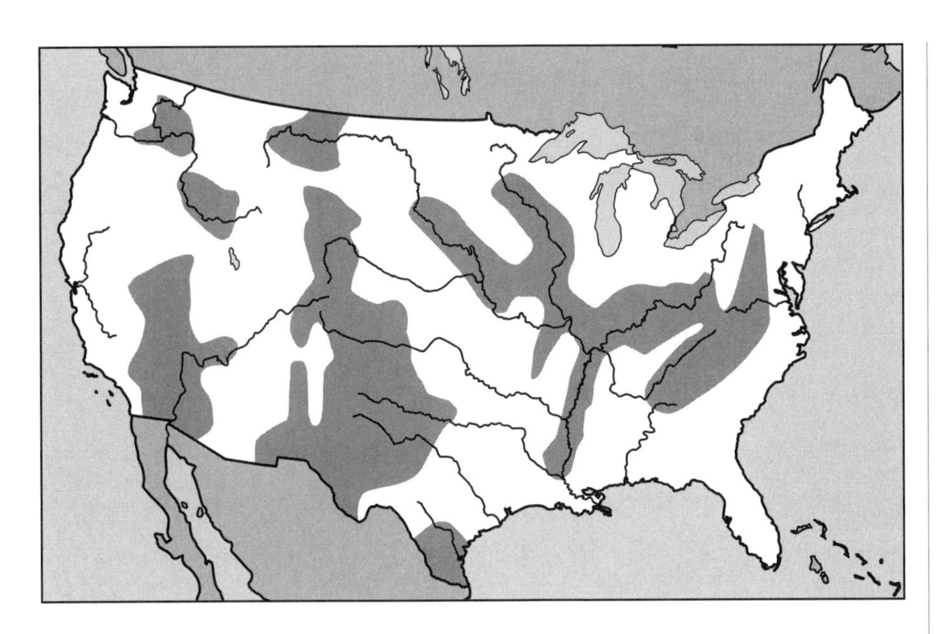

Figure 119 Areas affected by soil erosion in the United States.

restored. The world food production per capita could fall precipitously if the loss of topsoil continues unchecked.

The soil profile (Fig. 120 and Table 12) begins with the A zone, which contains most of the soil nutrients. It is a thin bed ranging from a few inches to a few feet thick, with a worldwide average of about 7 inches. Below this level lies the B zone, which is coarser and of poor soil quality. As the A zone thins and erosion brings the B zone to the surface, the potential for runoff and erosion increases because the poorer soil cannot sustain vegetation, whose roots hold the soil in place.

Soil erosion causes widespread degradation of the land surface. Falling rain erodes surface material by impact and runoff. The impact of raindrops striking the ground with a high velocity loosens material and splashes it up into the air. On hillsides, some of this material falls back lower down the slope. About 90 percent of the energy dissipates by the impact. Most of the impact splashes rise about 1 foot, with the lateral splash movement about four times the height.

Impact erosion is most effective in regions with little or no vegetative cover and subjected to sudden downpours such as desert areas. Splash erosion accounts for the puzzling removal of soil from hilltops where little runoff occurs. It also ruins soil by splashing up the light clay particles, which are carried away by runoff, leaving infertile sand and silt behind. Rainwater not infiltrating into the ground runs down the hillside and erodes the soil, cutting deep gullies into the terrain (Fig. 121).

Soil erosion rates vary depending on precipitation, the topography of the land, the steepness of slopes, the type of rock and soil materials, and the

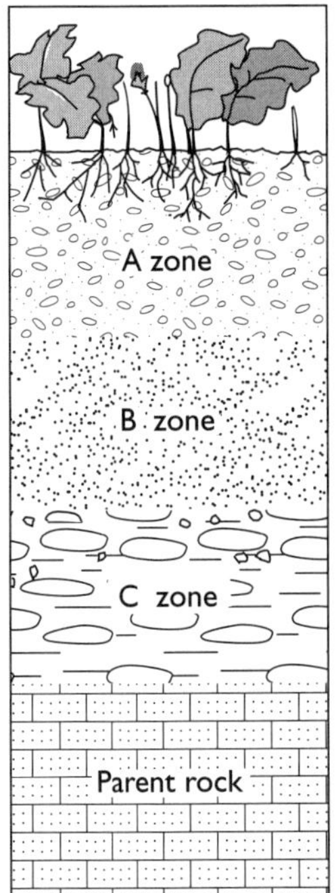

Figure 120 The soil profile. A zone—sand, silt, clay, and organic rich. B zone—sand, silt, clay, and organic poor. C zone—particles of parent rock plus material leached from above.

TABLE 12 SUMMARY OF SOIL TYPES

Climate	Temperate (humid) > 160 in. rainfall	Temperate (dry) < 160 in. rainfall	Tropical (heavy rainfall)	Arctic or desert
Vegetation	Forest	Grass and brush	Grass and trees	Almost none, no humus development
Typical area	Eastern U.S.	Western U.S.		
Soil type	Pedalfer	Pedocal	Laterite	
Topsoil	Sandy, light-colored; acid	Enriched in calcite; white color	Enriched in iron and aluminum, brick red color	No real soil forms because no organic material. Chemical weathering very low
Subsoil	Enriched in aluminum, iron, and clay; brown color	Enriched in calcite; white color	All other elements removed by leaching	
Remarks	Extreme development in conifer forest abundant humus makes groundwater acid. Soil light gray due to lack of iron	Caliche—name applied to accumulation of calcite	Apparently bacteria destroy humus, no acid available to remove iron	

Figure 121 *Gully erosion of topsoil, Umatilla County, Oregon.*

(Photo by K. N. Phillips, courtesy USDA Soil Conservation Service and USGS)

amount and type of vegetative cover. Efforts to increase worldwide crop production by deforestation, irrigation, artificial fertilizers, genetic engineering, and other scientific methods could ultimately fail if the topsoil erodes away. Three times as much food must be grown on the existing land to meet the global demands of a doubling of the human population projected for the middle of this century. Most of this increase must be made by new technology, especially since a significant portion of the world's cropland has already been lost.

To aggravate the problem, the world's rivers are becoming heavily sedimented from topsoil erosion. This is particularly true in Africa, which has the worst erosion in the world. In the United States, eroding cropland is costing nearly $1 billion annually because of polluted and sedimented rivers and lakes. The sediments also severely limit the life expectancy of dams built for water projects such as irrigation. Therefore, the most effective means to control silt buildup is to adopt effective soil-conservation measures in the watershed to limit topsoil lost by erosion.

Intensive agriculture has reduced by half the average soil depth in the United States over the last 150 years. In the 1980s, cropland in the United States shrank by 7 percent mostly due to rapid urbanization, with 1 percent or more of the most productive farmland lost annually. Expected rises in temperatures, increased evaporation rates, and changes in rainfall patterns by global warming could further weaken the nation's ability to grow sufficient food for its own consumption. The production of excess food for export might therefore be severely restricted, leading to mass starvation in countries that have already ruined their land and cannot adequately feed themselves without outside aid.

RIVER EROSION

The primary purpose of a river is to transport debris eroded at its headwaters and along its banks. The sediment originates from rocks weathered by wind, rain, and ice. At times, the river erodes its bed. At other times, the river builds it back up. Erosion and sedimentation therefore determine the shape of a river course from one confined to a single straight channel when eroding to one that meanders or is braided when clogged with debris. The sediment-laden stream eventually empties into a standing body of water, where sediments fall out of suspension.

Rivers erode valleys and provide a system of drainage delicately balanced with the climate, topography, and lithology. Individual streams and their valleys join into networks that display various types of drainage patterns, depending on the terrain. In areas of exposed bedrock, drainage patterns depend on the lithology of the underlying rocks, the attitude of rock units, and the arrangement and spacing of planes of weakness encountered by runoff.

Drainage patterns are also influenced by topographic relief and rock type. They provide important clues about the geologic structure of an area.

A river basin is the entire region from which a stream and its tributaries receive water. Each tributary is fed by smaller tributaries down to the smallest rill. Sediment grains loosened by raindrop impact erosion are carried downstream and dumped into the ocean. The products of erosion then settle onto the ocean floor, where they are consolidated by cementing agents such as calcium or silica and turned into sandstones, siltstones, and mudstones.

Rivers are the primary agents for transporting the products of erosion and play the most important role in carving out the landscape. Weathering, downslope movement, and river flow work together to reshape the continents. No matter how pervasive mountain ranges are, they eventually lose the battle with erosion and fall to the sea. Even in the most arid regions, the principal topographic features are a consequence of excavation by stream erosion. Streams carry to the sea sediments eroded from the highlands in the continental interiors. Along the way, they carve out new landforms, including deep ravines and winding valleys.

When a river captures a nearby stream, known as piracy, it creates a larger expanse of flowing water. The river grows at the expense of other streams and becomes dominant because it contains more water, erodes softer rocks, or descends a steeper slope. The river therefore has a faster headward erosion that undercuts the divide separating it from another stream and captures its water.

Figure 122 *Severe stream bank erosion along Muddy Creek, Cascade County, Montana.*

(Photo by T. McCabe, courtesy of USDA Soil Conservation Service)

Stream erosion (Fig. 122) deepens, lengthens, and widens valleys. At the head of a stream, where the slope is steep and water flow is fast, downcutting lengthens the valley by headward erosion, which is mainly how streams cut into the landscape. Farther downstream, both the velocity and discharge increase while the sediment size and the number of banks decrease, allowing the river to transport a greater load with lesser slope.

Erosion widens a stream valley by creep, landsliding, and lateral cutting. The process is most pronounced on the outsides of irregular curves where the valley side might be undercut by flowing water. Therefore, streams with migrating curves tend to widen their valleys. Many streams have distinctive symmetrical curves called meanders that uniformly distribute the river's energy.

Rivers erode by abrasion and solution. Abrasion occurs when the transported material scours the sides and bottom of the channel. The impact and drag of the water itself also erode and transport material. Most dissolved matter in a stream originates from groundwater draining from a breached water table. Materials such as limestone dissolve in river water that is slightly acidic. Limestone also acts as a buffer to maintain acidity levels within tolerable limits for aquatic life.

The rate of erosion in the drainage basin depends on the rainfall, evaporation, and vegetative cover. A river's ability to erode and transport material depends largely on the velocity, the water flow, the stream gradient, and the shape and roughness of the channel. The average rate of erosion in the United States is about 2.5 inches per 1,000 years. The Columbia River basin has the lowest erosion rate at 1.5 inches per 1,000 years, and the Colorado River basin has the highest at 6.5 inches per 1,000 years.

A river valley is a low-lying track of land traversed by a river or stream and bordered on both sides by higher ground called a floodplain. A narrow valley is not much wider than the river channel itself, whereas a broad valley exceeds many times the width of the river channel. A narrow valley is carved out by a fast flowing river that is actively downcutting in areas of regional uplift, called the youthful stage. Some narrow valleys slice through resistant rocks that slow the lateral cutting of a river and commonly display rapids and waterfalls.

A river widens its valley as it flows along a leveling grade and is no longer rapidly downcutting, called the mature stage. This condition occurs mostly near the mouth of a river, where wide floodplains exist. Meanders are common features of wide valleys, especially in areas with uniform banks composed of easily erodible sediments. The valley might widen by flooding, weathering, and mass wasting. Many river valleys were also widened by glaciers during the Ice Age, converting V-shaped valleys into U-shaped ones (Fig. 123).

The river velocity is determined by the roughness, shape, and curving of the channel along with the river slope. A typical river has a slope of several hundred feet per mile near its headwaters declining to just a few feet per mile near its mouth. For instance, the lower Mississippi River has a slope of less

Figure 123 *A U-shaped glaciated valley, Red Mountain Pass, south of Ouray, Colorado.*

(Photo by L. C. Huff, courtesy USGS)

than 6 inches per mile. A steep, rapidly flowing river such as the Colorado generally has a slope of 30 to 60 feet per mile and a slope angle of between one-third and two-thirds degrees.

WIND EROSION

Wind is the most important active agent of erosion, transportation, and deposition of sediments in desert regions. Deserts generate some of the strongest winds due to the effects of high-pressure systems and the rapid heating and cooling of the land surface. The winds produce sandstorms and dust storms that work together to cause wind erosion. Wind erosion develops mainly by the removal of large amounts of sediment during windstorms, forming a deflation basin. The wind often excavates hollowed-out areas called blowouts (Fig. 124), recognized by their typical concave shapes.

Wind erosion causes deflation and abrasion. Deflation is the removal of sand and dust particles by the wind. It usually occurs in arid regions and unvegetated areas such as deserts and dry lake beds. As smaller soil particles blow away during dust storms, the ground coarsens over time. The remaining sand tends to roll, creep, or bounce with the wind until it meets an obstacle, whereupon it settles and builds into a dune.

Abrasion, produced by wind-driven sand grains, causes erosion near the base of a cliff. When acting on boulders or pebbles, abrasion pits, etches,

Figure 124 *A wind blowout in Fremont County, Idaho, in August 1921.*

(Photo by H.T. Stearns, courtesy USGS)

grooves, and scours exposed rock surfaces. Maximum erosion effects occur during strong sandstorms, with sediment grains generally rising less than 2 feet above the ground. These abrasive effects occur most commonly on fence posts and power poles.

Grains of sand march across the desert floor under the influence of strong winds by a process known as saltation (Fig. 125). The sand grains become airborne, rising no more than a foot or two above the ground. When

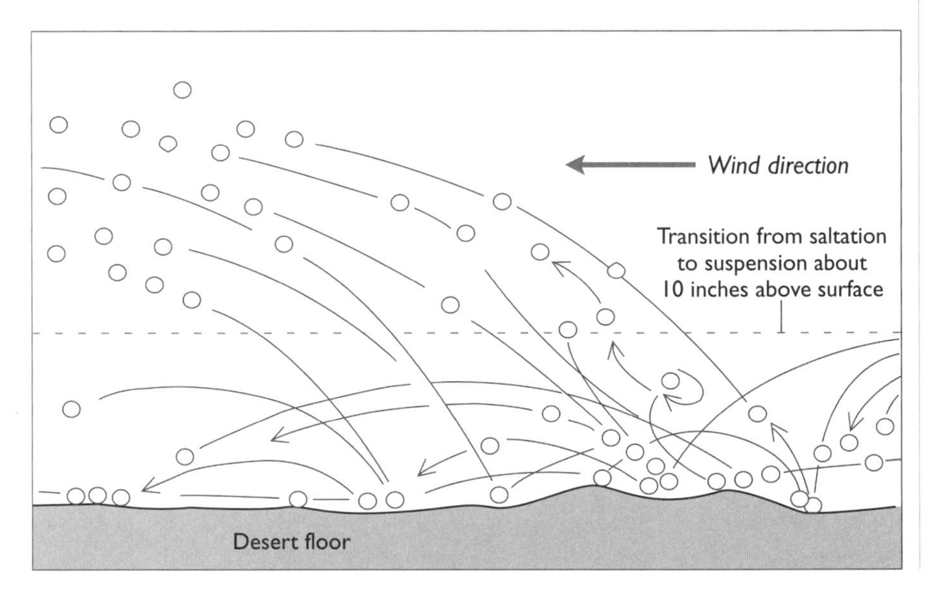

Figure 125 *The process of saltation causes sand to march across the desert floor.*

landing, they dislodge additional sand grains, which repeat the process. The rest of the moving sand travels forward along the ground by rolling and sliding. The constant motion of the sand abrades sediment grains, producing a frosted appearance.

Wind erosion often removes the fine material from the surface, leaving a layer of pebbles to prevent further erosion. Over a period of thousands of years, deserts develop a protective shield of pebbles varying in size from a pea to a walnut, which are too heavy for the strongest desert winds to pick up. Thus, the desert shield helps hold down sand grains and create a stable terrain. Any disturbance on the surface can spawn a new generation of roving sand dunes.

Arches often found in desert regions are spans of rock that have no natural stream flowing beneath them. They are formed when rock eroded at different rates due to a variance in resistance to erosional forces. They were created partly by wind erosion of thick sandstone beds. Rainwater first loosened the sand near the surface, while wind removed the loose sand grains. Wind erosion then abraded the rock, cutting through it in a manner similar to sandblasting.

GLACIAL EROSION

The effects of recent glacial events are readily seen in the landscape of the northern regions. Continental and alpine glaciers produced a variety of erosional and depositional landforms. Due to the variable types of forms and deposits, the environmental geology in a recently glaciated area can be highly complex. The wide variety of earth materials in previously glaciated areas requires special consideration during construction.

The unique glacial landscapes owe their existence to immense glaciers that swept down from the polar regions and destroyed everything in their paths during the ice ages. Thick sheets of ice overran continents, and alpine glaciers grew on nearly every mountain peak. Their legacy remains as deeply eroded rock in the high ranges of the world. The glaciers left an unusual collection of structures, including cirques, kettle holes, glacial lakes, flood-ruptured ground, and many other landforms sculpted by ice.

Thick deposits of glacial rocks called erratics, tillites, and moraines cover many northern lands. Glacial sediments buried older rocks, producing elongated hillocks called drumlins. Glacial debris from outwash streams formed long, sinuous sand deposits called eskers. Mounds of glacial sand piled up into kames. Finer sediment falling to the bottoms of glacially fed lakes created varves, whose distinct layering provides a convenient means of correlating the various glacial episodes.

Figure 126 *Glacial moraine on Eight Mile Creek, Lemhi Range west of Leadore, Lemhi County, Idaho.*

(Photo by W. C. Alden, courtesy USGS)

Glacially derived sediments covered large parts of the landscape, burying older rocks under thick layers of glacial till. The deposits are nonstratified material comprising clay and intermediate-sized boulders laid down directly by glacial ice. The simplest of all glacial landforms are moraines (Fig. 126). These are accumulations of rock material carried by a glacier and deposited in a regular, usually linear pattern resulting in a recognizable landform.

The basement rocks comprising the nuclei of the continents are exposed in broad, low-lying, domelike structures called shields. Many shields, such as the Canadian Shield covering most of eastern Canada and the Fennoscandian Shield in northern Europe, are fully exposed in areas ground down by flowing ice sheets during the last ice age. The soils in these regions are thin from glacial erosion, however, and would soon wear out with extensive agriculture.

EARTH MOVEMENTS

Slopes are the most common and among the most unstable landforms. Under favorable conditions, the ground can give way even on the gentlest slopes, contributing to the sculpture of the landscape (Fig. 127). Slopes are therefore inherently unstable and only temporary features over geologic time. The slope geometry along with the composition, texture, and structure of the soil determine the formation strength. Changes in pore pressure and water content can weaken the friction between rock layers. The maximum natural inclination of a slope, called the angle of repose, is self-regulating because slides bring the slope back to its critical state when it becomes over-

Figure 127 *Landslides near San Bautista, San Benito County, California.*

(Photo by R. D. Brown Jr., courtesy USGS)

steepened. Therefore, the amount of sediment that accumulates balances the amount carried away by landslides.

All earth movements such as landslides and related phenomena, including rockfalls, mudflows, earth flows, liquefaction, and subsidence, are naturally recurring events that have become increasingly hazardous because people continue crowding into susceptible lands. Earth materials are constantly on the move at rates varying from imperceptible creep of soil and rock to catastrophic landslides and rockfalls traveling at tremendous speeds that often result in death and destruction. In addition, ground failures occur when subsurface sediments liquefy during earthquakes or violent volcanic eruptions, inflicting considerable harm to people and their property.

Landslides (Fig. 128) are rapid downslope movements of soil and rock materials triggered mainly by seismic activity and severe storms. Single large slides in populated areas can cost tens of millions of dollars. In the United States, the direct yearly costs from damage to highways, buildings, and other facilities including indirect costs from the loss of productivity often exceed $1 billion. Single large slides in populated areas can cost tens of millions of dollars. Fortunately, landslides in this country have not resulted in a major loss of life as they have in other parts of the world because most catastrophic slope failures generally occur in sparsely populated areas.

The principal types of landslides are falls, topples, slides, spreads, and flows. Slides involving bedrock with strong, resistant rocks overlying weaker beds are called rock slides and slumps (Fig. 129). Material slides downward in a curved plane, tilting up the resistant unit, while the weaker rock flows out into a heap. Slides consisting of overburden alone are called debris slides and are the most dangerous slope movement with respect to human life.

The most damaging landslides in the United States occur in mountainous regions, including the Appalachian and Rocky Mountains and ranges along the Pacific coast. In California, landslides are pervasive and cause considerable property damage. Repeated heavy rains and floods often devastate hillsides, setting off landslides that destroy or seriously damage houses.

Massive landslides are often triggered by earthquakes. The size of the affected area depends on the earthquake magnitude, the geology and topography of the land, and the amplitude and duration of the ground motion. During the August 1959 Hebgen Lake, Montana, earthquake, a single large slide gouged a huge scar in the mountainside (Fig. 130). Debris traveled uphill on the opposite side of the valley, damming the Madison River and creating a large lake.

Figure 128 *A landslide that extends over the Pacific Coast Highway, the Pacific Palisades areas, Los Angeles County, California, on March 31, 1958.*

(Photo by J. T. McGill, courtesy USGS)

Figure 129 *Slumping of unstable earth materials undermines the foundations of houses in the Pacific Palisades area of southern California.*

(Photo by J. T. McGill, courtesy USGS)

Earthquake-induced landslides often result in wide area destruction. The 1971 San Fernando, California, earthquake produced nearly 1,000 slides distributed over 100 square miles of remote and hilly mountainous terrain. The 1976 Guatemala City earthquake triggered some 10,000 landslides throughout an area of 6,000 square miles. During the rainy season in Ecuador, an earthquake on March 5, 1987, shook loose fierce mud slides that buried villages in the rugged hilly region, killing more than 1,000 people.

Destructive landslides are also triggered by the removal of lateral support of a slope by erosion from streams, glaciers, waves, or ocean currents. The slides are initiated by previous slope failures and human action, such as excavation and other forms of construction work. The ground gives way under excess loading by the weight of rain, hail, or snow. In addition, the weight of buildings and other structures tends to overload a slope, causing it to fail.

Other common landslide-triggering mechanisms include explosions that break the bond holding the slope together, overloading the slope making it no longer able to support the new weight, undercutting at the base of the slope, and waterlogging sediments by rain or melting snow. Water increases the weight of the slope and decreases the internal cohesion of the overburden. The effect water has as a lubricant is limited, however, and mainly results in the loss of cohesion when filling the spaces between soil grains.

Just a minor difference in soil density determines whether a landslide becomes a fast-moving killer or merely one that slowly slumps downhill. The addition of water causes sudden slides that accelerate to great speeds. As the material collapses, water pressure increases in the pores between the grains in the soil. This reduces the friction between the grains, triggering a landslide. With increased soil density, the grains have to move apart to slide past each other, thereby increasing pore size. Water pressure decreases within the soil pores, which increases friction, resulting in a slump.

Avalanches (Fig. 131) are snow slides that usually begin with a mass of compacted snow resting on a steep bank of weaker snowpack. They are triggered by disturbances such as earthquakes, loud noises, or skiers. On January 16, 1995, a blizzard in the foothills of the Himalayas in Kashmir, northern India, stranded hundreds of people who abandoned their cars and buses on a one-lane highway to take shelter inside a 1.5-mile-long tunnel. Without

Figure 130 *The August 1959 Madison Canyon slide, Madison County, Montana.*

(Photo by J. R. Stacy, courtesy USGS)

Figure 131 *Avalanches across railroad and highway Anchorage district, Cook Inlet region from the March 27, 1964, Alaskan earthquake.*

(Photo courtesy USGS)

warning, an avalanche struck burying everything in the area. Some people managed to escape the tunnel before thousands of tons of snow completely closed it off. Several days later, bulldozers and villagers armed with shovels dug through the wall of snow, only to find the tunnel filled with frozen bodies.

Rockfalls or soil falls involve material dropping at the velocity of free fall from a nearly vertical mountain face. They are particularly hazardous to highways in mountainous terrain, especially after a heavy downpour (Fig. 132). Rockfalls range in size from individual blocks plunging down a mountain slope to the failure of huge masses of rock weighing hundreds of thousands of tons falling straight down a mountain slope. Individual blocks commonly come to rest at the base of a cliff, forming a loose pile of angular blocks called a talus cone.

If large blocks of rock drop into a standing body of water, such as a lake or fjord, immensely destructive waves are set into motion. A 1958 earthquake in Alaska triggered an enormous rockslide that fell into Lituya Bay, generating a gigantic wave that surged 1,700 feet up the mountainside. Trees toppled over like matchsticks when a massive surge of seawater inundated the shores. Coastal landslides of large magnitude can also generate destructive tsunamis. This hazard is particularly feared in Norway, where small deltas might provide the only available flat land at sea level. Waves generated by rockfalls can range from 20 to 300 feet high and cause considerable damage as they burst through local villages.

Figure 132 *A rockfall on Interstate 70, Jefferson County, Colorado, on May 8, 1973.*

(Photo by W. R. Hansen, courtesy USGS)

Rockslides are large and destructive earth movements, often comprising millions of tons of rock when a mass of bedrock fragments during the fall. The material behaves as a fluid. It spreads out on the floor below, often flowing some distance uphill on the opposite side of a valley. Rockslides are most prone to develop when areas of weakness, such as bedding planes or jointing, lie parallel to a slope, especially when undercut by erosion.

A rockslide southeast of Glacier Point in Yosemite National Park, California on July 10, 1996, sent 160,000 tons of granite that broke off a cliff plunging nearly 2,000 feet at more than 160 miles per hour. House-sized boulders bounced down the cliff face and tumbled to the floor of Yosemite Valley. The slide caused a hurricane-like air blast that leveled as many as 2,000 trees, some with their bark completely stripped off. Similarly, an earthquake in 1872 triggered a rockfall, which created an air blast that moved a house a couple of inches off its foundation.

The air blast represents a poorly understood collateral hazard of rockfalls. It is similar to the effect of dropping a book parallel to the ground, which forces the air out from underneath it. Apparently, a falling mass of rock must remain together long enough to shove air out of its way. Geologists might have to reassess hazard zones marked on maps of Yosemite and other mountainous areas to take into account the danger from air blasts.

Some of the largest and most damaging slides occur in the ocean. The constant tumbling of seafloor sediments down steep banks churn the ocean bottom into a murky mire. During the 1964 Good Friday Alaskan earthquake, submarine slides carried away large sections of the port facilities at Whittier, Valdez, and Seward (Fig. 133). Submarine flow failures can generate large tsunamis that overrun parts of the coast. For example, in 1929, an earthquake on the coast of Newfoundland set off a large undersea slide that triggered a tsunami, killing 27 people. On July 3, 1992, what appeared to be a large undersea slide sent a 25-mile-long, 18-foot-high wave crashing down onto Daytona Beach, Florida, overturning automobiles and injuring 75 people.

On July 17, 1998, a train of three giant waves 50 feet high swept away 2,200 residents of Papua New Guinea. The disaster was originally blamed on a nearby undersea earthquake of 7.1 magnitude, but it was considered much too small to heave up waves to such heights. Evidence collected during marine surveys of the coast implicated a submarine slide or slump of underwater sediment large enough to spawn the waves. The continental slope bears a thick carpet of sediments, which in places has slid downhill in rapid landslides and slower-moving slumps. The evidence on the ocean floor suggests that large tsunamis can be generated by moderate earthquakes when accompanied by landsliding. This phenomenon makes the hazard much more dangerous than was once thought.

Figure 133 A railroad yard and warehouse damaged at Seward due to submarine slides from the March 27, 1964, Alaskan earthquake.

(Photo courtesy USGS)

Coastal slides carve out deep submarine canyons in continental slopes. The slides consist of sediment-laden water much denser than the surrounding seawater, allowing sediments to move swiftly along the ocean floor. These muddy waters, called turbidity currents, can move down the gentlest slopes and transport immensely large blocks. Turbidity currents are also initiated by river discharge, coastal storms, or other currents. They deposit huge amounts of sediment that build up the continental slopes and the smooth ocean bottom below.

The continental slopes incline as much as 60 to 70 degrees and plunge downward for thousands of feet. Sediments that reach the edge of the continental shelf slide off the continental slope by the pull of gravity. Huge masses of sediment cascade down the continental slope by gravity slides that can gouge out steep submarine canyons and deposit great heaps of sediment. They are often as catastrophic as terrestrial slides and can move massive quantities of sediment downslope in a matter of hours.

The submerged deposits near the base of the main island of Hawaii rank among the largest landslides on Earth. On the southeast coast of Hawaii, on Kilauea Volcano's south flank, about 1,200 cubic miles of rock are slumping toward the sea at a speed of 4 inches or more per year (Fig. 134). It is the biggest thing on Earth that is moving in this fashion. Six miles

Figure 134 *A stepped topography produced by subsidence of large landslide blocks on the south flank of Kilauea Volcano, Hawaii.*

(Photo courtesy USGS/HVO)

below the volcano lies a nearly horizontal fault that is slipping at a rate of 10 inches per year, making it the fastest-moving fault in the world. Ultimately, some sort of failure will occur, far more destructive than any of the volcano's eruptions.

Gigantic slices of volcanoes have broken off the Hawaiian Islands and skidded across the ocean bottom, sometimes creating towering tsunamis that break on nearby shores. On the island of Kauai, a volcano built up the western portion of the island and then collapsed along its eastern edge in a giant landslide. Later, a new volcano grew in its place, only to collapse as well. Remnants of these massive landslides litter the seafloor around the island.

The largest example of an undersea rock slide from a Hawaiian volcano measured roughly 1,000 cubic miles in size and spread some 125 miles from its point of origin. The collapse of the island of Oahu sent debris 150 miles across the deep-ocean floor, churning the sea into gargantuan waves. When part of Mauna Loa collapsed and fell into the sea around 100,000 years ago, it created a tsunami 1,200 feet high that was not only catastrophic for Hawaii but might have even caused damage along the coast of California.

GROUND FAILURES

Liquefaction causes ground failures in water-saturated subterranean sediments during severe ground shaking accompanying earthquakes and violent volcanic eruptions. Generally, the younger and looser the sediment and the shallower the water table, the more susceptible the soil is to liquefaction. Sand boils occur when earthquakes turn a solid, water-saturated bed of sand underlying less permeable surface layers into a pool of pressurized liquid that spouts to the surface by artesianlike water pressures developed during the liquefaction process (Fig. 135). Sand boils often cause localized flooding and the accumulation of large deposits of sediment.

Ground failures associated with liquefaction include lateral spreads, flow failures, and loss of bearing strength. Lateral spreads are the horizontal movement of large blocks of soil generally on gentle slopes caused by the liquefaction of a subsurface layer during earthquakes. Horizontal movements on lateral spreads are up to 10 or more feet. However, where slopes are particularly favorable and the duration of the temblor is long, the ground movement might extend several times farther. Lateral spreads usually break up internally, forming numerous fissures and scarps.

The most catastrophic type of ground failure resulting from liquefaction are flow failures, involving liquefied soil alone or blocks of intact material riding on a layer of liquefied soil. These failures commonly move several tens of feet. However, under certain geographic conditions, they can travel tens of miles at speeds of many tens of miles per hour. Flow failures usually form in loose, saturated sands or silts on steep slopes. They originate both on land and under the sea.

Most clays temporarily lose strength when disturbed by earthquakes and behave as viscous fluids. If the loss of strength is significant, some clays called quick clays can fail catastrophically. Quick clay is composed primarily of flakes of clay minerals arranged in very fine layers, with a water content often exceeding 50 percent. Ordinarily, quick clay is a solid that can support more

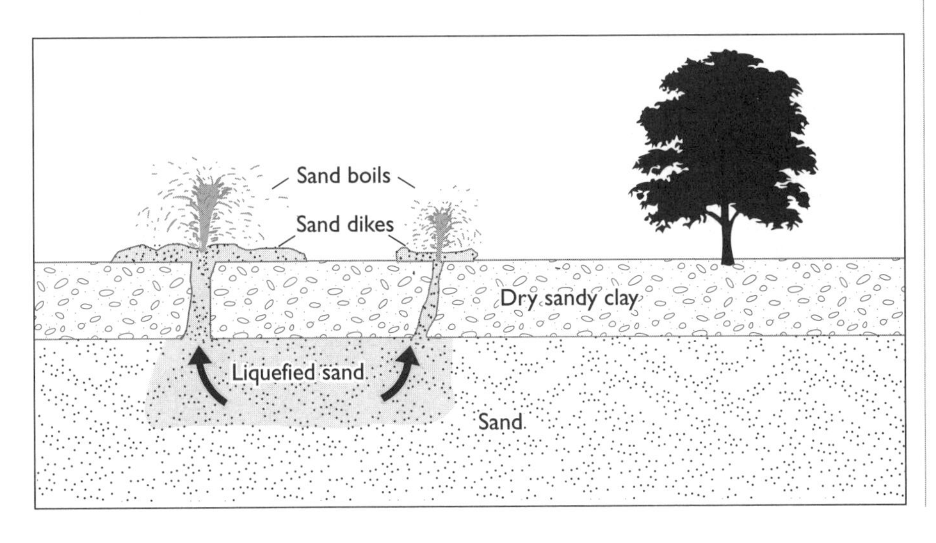

Figure 135 Sand boils are fountains of water and sediment that spout from a pressurized, fluidized zone during earthquakes.

than 1 ton per square foot of surface area. However, the slightest jarring motion from an earthquake can turn it into a fluid.

During the 1964 Alaskan earthquake, five large landslides devastated parts of Anchorage (Fig. 136) due to catastrophic failures of clays sensitive to ground motions. Many houses were destroyed when 30 city blocks underlain by a slippery clay substratum slid toward the sea. The slides resulted from the failure of layers of quick clay along with other beds composed of saturated sand and silt. Because of the severity of the earthquake, the clay layers lost strength while the sand and silt layers were subjected to liquefaction.

When the soil supporting buildings or other structures liquefies and loses bearing strength, large deformations can occur within the soil, causing structures to settle or subside (Fig. 137). Soils that liquefy beneath buildings distort the general subsurface geometry. This results in bearing failures and subsidence that can cause a building to tilt. Normally, these deformations occur when a layer of saturated, cohesionless sand or silt extends from near the surface to a depth equal to the width of the building. The most spectacular

Figure 136 *Collapse of Fourth Avenue in Anchorage due to a landslide caused by the March 27, 1964, Alaskan earthquake.*

(Photo courtesy Alaskan Earthquake USGS)

Figure 137 *Highway 1 bridge destroyed by liquefaction of river deposits at Struve Slough during the October 17, 1989, Loma Prieta, California, earthquake.*

(Photo by G. Plafker, courtesy USGS)

example of this type of ground failure occurred during the June 16, 1964, Niigata, Japan, earthquake, when several four-story apartment buildings tilted as much as 60 degrees (Fig. 138). They were later jacked upright and underpinned with pilings to prevent a recurrence of ground failure.

Figure 138 *Apartment buildings in Niigata, Japan, that tipped because of the loss of bearing strength caused by liquefaction in the underlying sediments during the 1964 Niigata earthquake.*

(Photo courtesy USGS)

Earthquakes often trigger soil slides in weakly cemented, fine-grained materials that form steep, stable slopes. Soil on steep hillsides suddenly transforms into a wave of sediment, sweeping downward at speeds of more than 30 miles per hour. Precipitation frees dirt and rocks by increasing the pore water pressure within the soil. As the water table rises and pore pressure increases, friction holding the topsoil layer to the hillside decreases and is overcome by the pull of gravity.

The slow downslope movement of soil, called creep (Fig. 139), is often recognized by poles, fence posts, and trees tilting downhill. Creep is a more rapid movement of near-surface soil material than the sediment below and is particularly rapid where frost action is prominent. After a freeze-thaw sequence, material moves downslope by the expansion and contraction of the ground. Under these unstable slope conditions, trees are unable to take root. Only grasses and shrubs can grow on the hillsides. If creep is especially slow, tree trunks bend downhill, while new growth attempts to straighten them. If the creep is continuous, trees lean downhill in their lower parts and become progressively straighter higher up.

With additional water content, the weight of the overburden increases and the stability of the slope decreases as the adhesion of grain particles is reduced, resulting in an earth flow (Fig. 140). Earth flows are a more visible

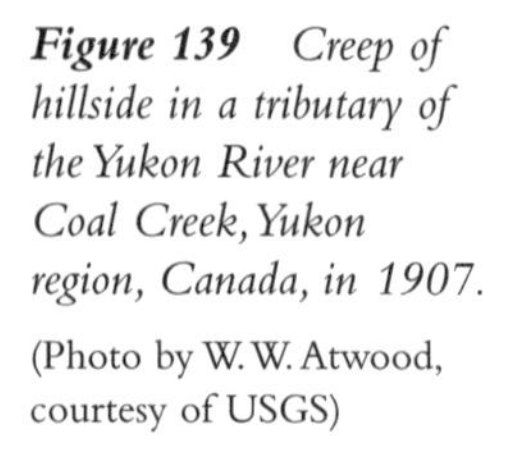

Figure 139 *Creep of hillside in a tributary of the Yukon River near Coal Creek, Yukon region, Canada, in 1907.*

(Photo by W. W. Atwood, courtesy of USGS)

Figure 140 *An ancient slump and earth flow exposed by recent landsliding, Fort Spokane area, Washington.*

(Photo by F. O. Jones, courtesy USGS)

form of movement and characterized by grass-covered, soil-blanketed hills. Although generally minor features, earth flows can grow considerably large, covering several acres. They usually have a spoon-shaped sliding surface upon which a tongue of overburden breaks away and flows for a short distance, forming a curved scarp at the point where material separates from the hillside.

Sediments that swell or shrink as the moisture content changes are called expensive soils. Swelling soils produce the greatest yearly property losses of all geological hazards. The parent materials for expansive soils are derived from the decomposition of volcanic and sedimentary rocks into clay minerals that tend to form highly unstable slopes. Expansive soils are abundant in geologic formations in the Pacific Coast, the Rocky Mountain region, the Basin and Range Province, the Great Plains, the Gulf Coastal Plain, and the lower Mississippi River Valley. Damages to buildings and other structures built on expansive soils cost the United States several billion dollars annually.

Heavy runoff in mountainous regions forms rapidly moving sheets of water that pick up large quantities of loose material, resulting in mudflows that can cause considerable damage (Fig. 141). The floodwaters flow into a stream, where the muddy material suddenly concentrates in the stream channel. The dry streambed rapidly transforms into a flash flood that moves swiftly downhill, often with a steep, wall-like front. Mudflows behave as a viscous fluid and

Figure 141 *A large mudflow of February 2, 1953, on the Nespelem River—Omak Lake Valley area, Okanogan County, Washington.*

(Photo by F. O. Jones, courtesy USGS)

often carry a tumbling mass of rocks and large boulders. Heavy rains falling on loose pyroclastic material on the flanks of volcanoes also produce mudflows.

Among the most extensive geologic activities caused by human action is the sinking of the ground by overpumping subterranean fluids. Subsidence is the settling or collapse of the land surface either locally or over broad regional areas without appreciable horizontal movement. This is due mostly to the withdrawal of underground fluids or by shock waves from earthquakes. Earthquake-induced subsidence in the United States has occurred mainly in California, Hawaii, and Alaska. The subsidence results from vertical displacements along faults that can affect broad areas.

Subsidence is a growing problem that is worsening as people continue to draw heavily on groundwater and petroleum. Many parts of the world have been steadily sinking due to the withdrawal of large quantities of groundwater. Generally, subsidence is roughly 1 foot for every 20 to 30 feet of lowered water table. Underground fluids fill intergranular spaces and support sediment grains. The removal of large volumes of fluid results in a loss of grain support, a reduction of intergranular void spaces, and the compaction of clays, which causes the land surface to subside wherever widespread subsurface compaction occurs.

Subsidence by the withdrawal of groundwater can produce fissures or open cracks in the ground. Subsidence can also cause the renewal of surface movement in areas cut by faults. Surface fissuring and faulting resulting from the withdrawal of groundwater is a potential problem near Las Vegas, Nevada, and in the arid regions of California, Arizona, New Mexico, and Texas.

The most dramatic examples of subsidence have occurred along the Gulf coast of Texas as well as in California and Arizona. Intense pumping of groundwater for agricultural purposes has caused large areas of California's San Joaquin Valley to subside. The arid region is so dependent on groundwater that it accounts for about one-fifth of all well water pumped in the United States. The ground has been sinking at rates of up to 1 foot a year. In the northern part of the valley, subsidence has dropped the land surface more than 10 feet below sea level, requiring protective dikes to prevent flooding. Subsidence in some coastal areas has also increased susceptibility to flooding during earthquakes or severe coastal storms.

Significant subsidence also occurs when water is added to sediments. This is especially true in the heavily irrigated, dry, western United States, where the land surface has lowered 3 to 6 feet on average and as much as 15 feet in the most extreme cases. The settling results when dry surface or subsurface deposits are extensively wetted for the first time since their deposition after the last ice age. The wetting causes a reduction in the cohesion between sediment grains, which move and fill in intergranular openings. The compaction produces an uneven land surface, resulting in depressions, cracks, and wavy surfaces. At other times, the land surface settles slowly and irregularly.

The collapse of abandoned underground coal mines, especially in the eastern United States, often leaves the strata above the mine workings with inadequate support, resulting in depressions and pits on the surface. In situ (operating in place) coal gasification and oil shale retorting can also cause the overlying ground to subside. Solution mining, using large volumes of water pumped into the ground to remove soluble minerals such as salt, gypsum, and potash, excavate huge underground cavities that can collapse and cause surface subsidence. When mining occurs under towns, the overlying buildings can be heavily damaged or destroyed.

The collapse of land overlying limestone caverns forms sinkholes 100 feet or more deep and up to several hundred feet across. The subsidence can cause extensive damage to buildings and other structures located over the pits formed by dissolving soluble minerals. Limestone and other soluble materials underlie large portions of the world. When groundwater percolates downward through these formations, it dissolves minerals, forming cavities or caverns. When the land overlying this cavern collapses, it forms a deep sinkhole.

One of the most dramatic examples of ground collapse occurred in Barton, Florida, on May 22, 1967, when a sinkhole 520 feet long and 125 feet

Figure 142 *A sinkhole 520 feet long, 125 feet wide, and 60 feet deep that collapsed under a house in Barton, Florida.*
(Photo courtesy USGS)

wide suddenly opened under a house (Fig. 142). On December 12, 1995, heavy rainfall and a sewer pipe break in San Francisco, California, created a huge sinkhole as deep as a 10-story building that swallowed a million-dollar house and threatened dozens more. Although the formation of sinkholes is a natural phenomenon, the process is accelerated by the withdrawal of groundwater or the disposal of wastewater into the ground.

Severe subsidence also occurs in permafrost regions. Solifluction is the slow downslope movement of waterlogged sediments that causes ground failures in colder climates. When frozen ground melts from the top down, during spring in the temperate regions or summer in permafrost regions, it causes the soil to glide downslope over a frozen base. Solifluction can create many construction problems, especially in areas of permafrost. Foundations must extend down to the permanently frozen layers, or entire buildings might be carried off.

After discussing erosion, landslides and related phenomena, along with ground collapse, the next chapter takes a look at the geologic hazards of desert regions.

8

DESERTIFICATION

DESERTS AND DROUGHTS

This chapter examines the effects of droughts, advancing deserts, and sand in motion. Droughts are periods of abnormally dry weather resulting from shifting precipitation patterns around the world. They can be sufficiently prolonged for the lack of water to undermine agriculture (Fig. 143). Global warming can potentially increase the frequency and severity of droughts, with continental interiors that experience occasional droughts becoming permanently dry wastelands.

Monsoon rains that bring life-sustaining water to half the people of the world could be significantly curtailed during a period of global warming. The changing climate could make deserts out of once productive farmlands, limiting the world's ability to feed itself. Furthermore, strong winds blowing across deserts and arid lands would produce powerful dust storms that cause severe soil erosion and bands of roving sand dunes.

THE WORLD'S DESERTS

About one-third of Earth's landmass, or roughly 20 million square miles, is desert (Fig. 144 and Table 13). The arid lands are the hottest and driest

Figure 143 *An ear of corn that never reached maturity due to severe drought in the American South during the summer of 1986.*

(Photo by June Davidek, courtesy USDA)

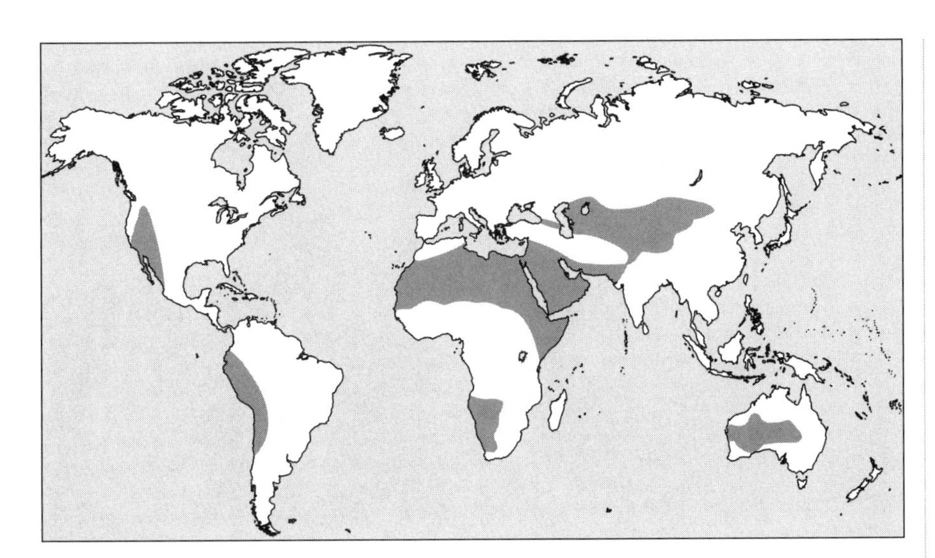

Figure 144 Location of worldwide deserts.

regions. Desert wastelands received only minor precipitation during certain seasons, while some areas have been essentially rainless for years. Only the hardiest plant and animal species with very unusual adaptations can tolerate these arid conditions. Often, when the rains arrive, heavy downpours cause severe flash floods that sweep away massive quantities of sediment and debris.

TABLE 13 MAJOR DESERTS

Desert	Location	Type	Area (Square Miles X 1,000)
Sahara	North Africa	Tropical	3,500
Australian	Western/interior	Tropical	1,300
Arabian	Arabian Peninsula	Tropical	1,000
Turkestan	Central Asia	Continental	750
North America	S.W. U.S./N. Mexico	Continental	500
Patagonian	Argentina	Continental	260
Thar	India/Pakistan	Tropical	230
Kalahari	S.W. Africa	Littoral	220
Gobi	Mongolia/China	Continental	200
Takla Makan	Sinkiang, China	Continental	200
Iranian	Iran/Afghanistan	Tropical	150
Atacama	Peru/Chile	Littoral	140

Figure 145 *A rain shadow zone is produced on the lee side of a mountain range, causing deserts in many parts of the world.*

Gigantic dust storms and sandstorms prevalent in desert regions also play a major role in shaping the arid landscape.

Most of the world's deserts lie in the subtropics in a broad band between 15 and 40 degrees latitude on either side of the equator. High precipitation levels in the Tropics leave little moisture for the subtropics, where the dry air cools and sinks. This produces semipermanent, high-pressure zones called blocking highs because they tend to block advancing weather systems from entering the region. Mountains also block weather systems by forcing rain clouds to rise and precipitate on the windward side of the range. The lack of precipitation on the leeward side or opposite end of the mountains results in a rain deficit, called a rain shadow zone (Fig. 145). This creates deserts such as those in the southwestern United States. Moist winds from the Pacific Ocean cool and precipitate as they rise over the Sierra Nevada and other mountain ranges in California, leaving regions to the east parched and dry.

Deserts are the most barren environments and lack significant plant and animal life. The forbidding Sahara Desert in northern Africa and the great central desert of Australia are among the least densely populated regions in the world. And the drylands bordering the deserts cover one-quarter of the landmass and support about 15 percent of the human population.

Deserts are also among the most dynamic of landscapes. They are constantly changed by blowing sand and drifting dunes. Gigantic dust storms and sandstorms prevalent in desert regions play a major role in shaping the arid landscape. Powerful sandstorms clog the skies with thousands of tons of sediment. Roving sand dunes driven across the desert by strong winds engulf everything in their paths. Coastal deserts are unique because they are areas where the seas meet the desert sands. The Namib Desert along the coast of Namibia, Africa, is the world's largest coastal desert. Its linear sand dunes, which are among the highest in the world, reaching 500 feet in height, are clearly visible from orbiting spacecraft (Fig. 146).

Desert sands are generally light colored and therefore have a high albedo, which is the ability of objects to reflect sunlight. The sand absorbs heat during the day, while the surface scorches at temperatures often

Figure 146 *Linear dunes in the northern part the Namib Desert, Namibia.*

(Photo by E. D. McKee, courtesy USGS)

exceeding 65 degrees Celsius. However, because the skies are generally clear at night, the thermal energy trapped in the sand quickly escapes due to its low heat capacity. This makes desert regions among the coldest environments. Even summertime temperatures at higher elevations can drop to near freezing at night. As a result, deserts have the highest temperature extremes of any environment.

Deserts typically receive less than 10 inches of average annual rainfall. Evaporation usually exceeds precipitation throughout the year. Only minor amounts of rain precipitate during certain seasons in the world's desert wastelands. Some regions, such as Egypt's Western Desert, have gone essentially without rain for many years. Because of these stringent conditions, desert areas cannot support significant human populations without artificial water supplies.

When the rains come to the desert, they are often violent and local, causing severe flash floods. A typical desert rain falls as a short torrential downpour that floods several square miles, then leaves the ground high and dry. While the flood is in motion, water levels in dry wadis rise rapidly and fall almost as fast, as the flood wave flows through the desert. Eventually, the floodwaters empty into shallow lakes that later dry up, or they soak into the dry, parched ground. Then for months or even years afterward, no rain will fall.

Deserts are home to some of the hardiest species on Earth, including plants whose seeds can survive a 50-year drought and rodents that spend their entire lives without taking a single drink of water. They instead survive solely off the water generated by their body's metabolism. Plants and animals use a variety of adaptations to survive desert conditions. They generally rely on the conservation of water and suspended animation during the driest part of the season.

The giant saguaro cactus (Fig. 147), common in the Sonora Desert of northwest Mexico and southwest United States, stores water in its trunk. Other plants extract moisture such as morning dew directly from the air. During the hottest part of the day, many animals retreat to underground burrows, where the temperature difference is significant. Even the space a few feet above the ground drops several degrees. Animals perch on small bushes to take advantage of the markedly cooler air.

During the short rainy season, aquatic species such as fish and amphibians must quickly lay their eggs before the ponds they briefly inhabit dry out. The animals then burrow into the bottom mud and lie dormant until the rains return. Australian desert frogs gorge themselves with water and burrow down as much as 3 feet, where they lie in suspended animation for months on end. During the next rainy season, the animals revive, their eggs hatch, and the cycle begins anew.

Figure 147 *Saguaro and other desert vegetation in the Sonora Desert, Arizona.*

(Photo by W. T. Lee, courtesy USGS)

Lungfish living in African swamps that seasonally dry out hole up for long stretches until the rains return. They burrow into the moist mud, leaving an air hole leading to the surface. They live in suspended animation, breathing with primitive lungs. Thus, they can survive out of water for several months or even a year or more if necessary. When the rainy season returns and the pond refills, the fish revive, breathing normally with gills.

The Namib Desert is inhabited by tiny fairy shrimps, whose eggs lie dormant for as long as 20 years or more. After a rare rain shower fills the dry, shallow basins, they subsequently become teeming with life. The shrimps must lay their eggs before the water evaporates in the hot sun, leaving the pools once again cracked and parched.

Perhaps the most impoverished desert on Earth is surprisingly in Antarctica. Its dry valleys (Fig. 148) host only meager signs of life, including blue-green algae on the bottoms of small glacier-fed lakes, soil bacteria, and a giant wingless fly. Antarctica has just two flowering plant species, which have undergone population explosions recently, possibly due to a warming climate. Delicate mosses and lichens if disturbed take a century to recover. The discovery of lichens in tiny pores on the undersides of rocks has spawned speculation that similar life-forms might inhabit the planet Mars, whose frigid terrain has many similarities to the frozen continent of Antarctica.

Figure 148 *Dry valleys and mountains in Antarctica.*

(Photo by F. R. Bair, courtesy U.S. Navy)

DROUGHT-PRONE REGIONS

Droughts are commonplace throughout the world. However, their effects are steadily worsening because of deepening poverty, increasing population, and the abuse of the land. Changes in land use are altering the hydrologic cycle, causing a permanent decrease in rainfall and soil moisture. During severe drought, not only do people die in tragically large numbers but livestock also perish, further reducing the food supply. Yet no definite evidence shows that the climatological mechanisms associated with droughts, floods, and tropical storms are changing. Instead, the effects of natural disasters are worsening by poverty, environmental damage, and rapid population growth.

Primitive farming techniques have devastated the land, causing a serious decline in agriculture. Under increasing pressure for food production, normally fallow fields are forced into production, which quickly wears out the soil. Most poor farmers cannot afford chemical fertilizers. The animal dung once used to enrich the soil is instead burned for fuel because forests have been cleared and supplies of firewood have dwindled. Moreover, deforestation causes the soil to lose much of its capacity to retain moisture, thereby reducing productivity and resistance to drought. The root cause of the agricultural crisis is growing population, soil erosion, and desertification. Therefore, famine is becoming more of a man-made disaster.

The Sahara Desert (Fig. 149), covering an area of 3.5 million square miles, about the size of the United States, is the largest arid region on Earth.

The sands of the Sahara Desert are advancing inexorably, engulfing the adjacent lands. Central Africa has lost much of its grazing land to the encroachment of the Sahara Desert, the southern extent of which crept 80 miles farther south between 1980 and 1990. The Sahel region south of the Sahara (Fig. 150) is a vast belt of drought that spreads across the African continent, parching the land and starving its inhabitants.

Desertification in the Sahel has been accelerating alarmingly. The process began more than 1,000 years ago when nomads of the Sahel lived as herders

Figure 149 *Linear dunes crested with barchanoid ridges in the northwest Sahara Desert, Algeria, northern Africa.* (Photo by E. D. McKee, courtesy USGS and NASA)

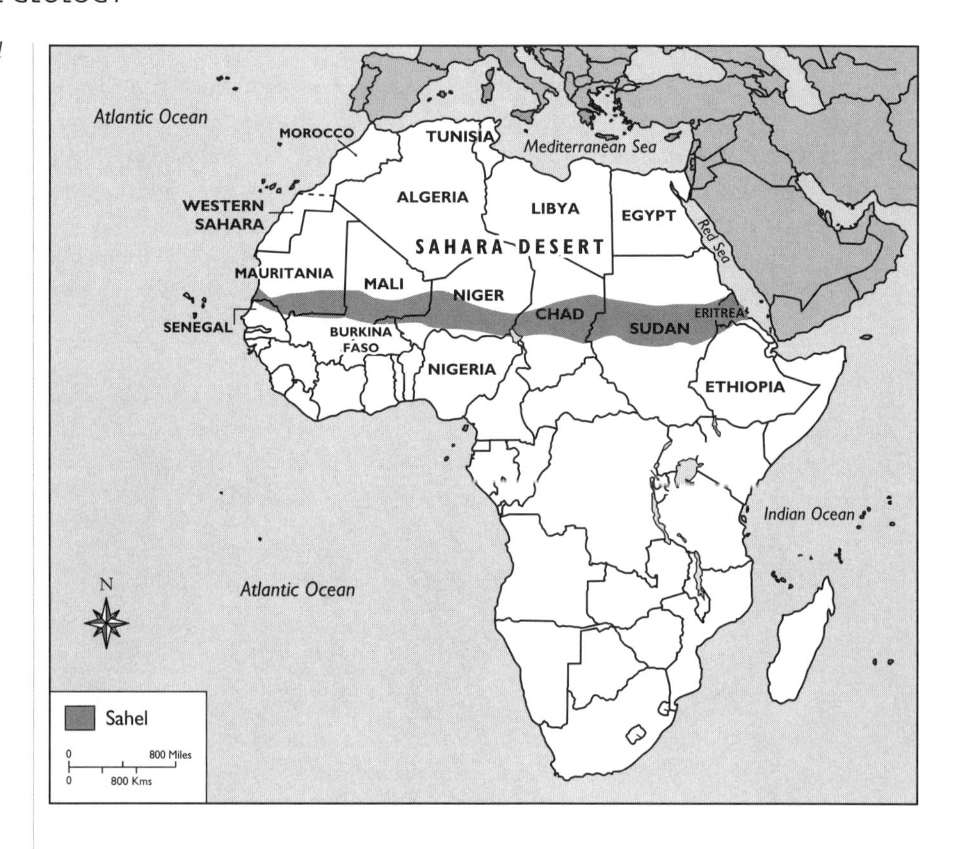

Figure 150 *The Sahel region of central Africa is an area of widespread drought, which is often overrun by the Sahara Desert.*

and hunters. They cut trees and set fires to improve grazing, converting the tropical forests into grasslands. The colonial era of the 19th century halted the nomadic life, forcing the people of the Sahel to settle the countryside and began farming and ranching. Cattle overgrazing further destroyed the already weakened soil, and desertification accelerated.

Droughts commonly occur in Africa, but their frequency and extent seem to be increasing. The 1984 African drought, whose resulting famine killed about 500,000 people, was predicted two years before the event. Unfortunately, the warnings were mostly ignored. To prevent the recurrence of this tragedy, researchers are using satellites to map vegetation across the entire continent. On the edges of deserts, satellite images chronicle where the grassland is disappearing and determine the amount of stress that vegetation experiences during a drought. They can depict how much vegetative cover is lost from one year to the next and ultimately help answer the larger ecological questions about vanishing forests and burgeoning human populations.

CAUSES OF DROUGHT

Droughts result from a shift in global precipitation patterns throughout the world (Fig. 151). Since Earth's total heat budget does not change significantly from year to year, regions that become unusually dry are matched, to some extent, by areas that become exceptionally wet. For example, during the 1980s, the United States endured a series of bad droughts, and Australia had its most severe drought in more than 100 years. An equally intense drought caused food shortages in southern Africa and seriously affected West Africa and the Sahel region of central Africa as well.

The droughts might have been triggered by an unusually warm tropical Pacific during an El Niño event and its accompanying atmospheric changes. The Pacific warming often steals rain from Australia, Indonesia, parts of Brazil, and eastern and southern Africa while flooding the normally dry west coast of South America. The El Niño of 1997, one of the worst on record, drenched western South America and eastern Africa while drying out Indonesia and southern Africa.

El Niños are anomalous warming conditions in the eastern equatorial Pacific Ocean due to the failure of the westerlies, the eastward-blowing trade winds. Modern El Niños, which return every few years, began about 5,000 years ago, soon after global sea levels steadied after melting of the ice sheets following the last ice age. Prior to this time, they came once every 15 to 70 years. The higher frequency of today's El Niño events appears to be a symptom of greenhouse gas pollution and global warming. By contrast, a colder than normal tropical Pacific known as La Niña might lead to higher than normal rainfall levels.

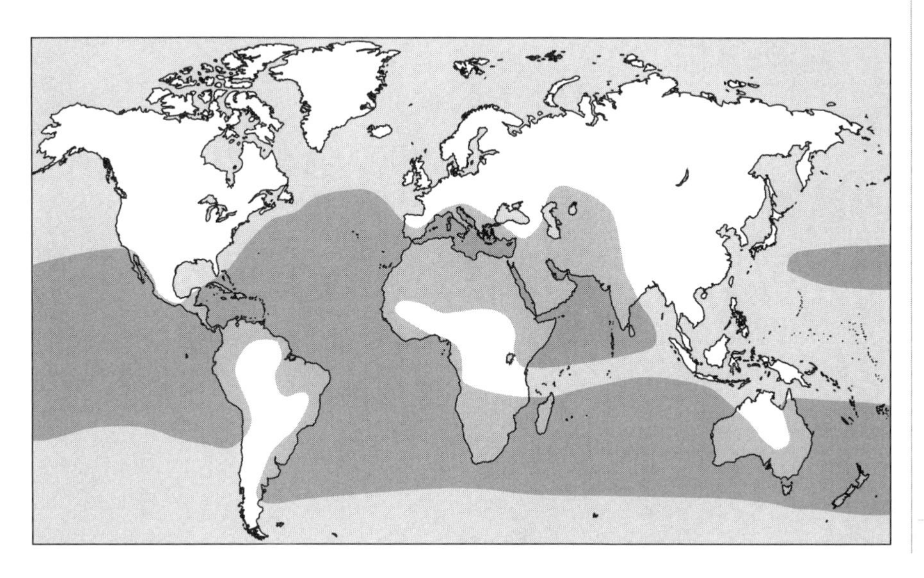

Figure 151 The global precipitation-evaporation balance. In the shaded areas, evaporation exceeds precipitation.

Droughts in the Sahel region seem to be associated with warmer-than-normal water in the Atlantic Ocean off West Africa. Unusually colder water appears simultaneously across the Atlantic as far north as the Caribbean and apparently influences precipitation. Changes in ocean currents induced by atmospheric circulation redistribute heat in the Atlantic to create the abnormal sea surface temperature pattern. The drought-related sea surface pattern develops in the months before the crucial summer rainy season. The precipitation activity is then shifted hundreds of miles to the south, reducing the moisture-laden winds reaching into West Africa and the Sahel region.

The seasonal winds called monsoons (Fig. 152) bring life-sustaining water to half the people of the world. The monsoon of southern Asia is possibly the most impressive seasonal phenomenon of the Tropics. The term monsoon, from the Arabic word *mausim,* meaning "season," applies to the wind system of the Arabian Sea that blows from the southwest during half the year and from the northeast during the other half. Generally, the term has come to signify any annual climatic cycle with seasonal wind reversals. The largest and most vigorous monsoons occur on the continents of Asia, Africa, and Australia.

The monsoons are seasonal changes in wind direction, alternately producing wet summers and dry winters. During the rainy season, periods of drenching squalls are interspersed with equal intervals of sunny weather lasting a week or more. During the monsoon's dormant phase, the weather is hot, dry, and stable, with an absence of tropical storms. Variations in annual precipitation

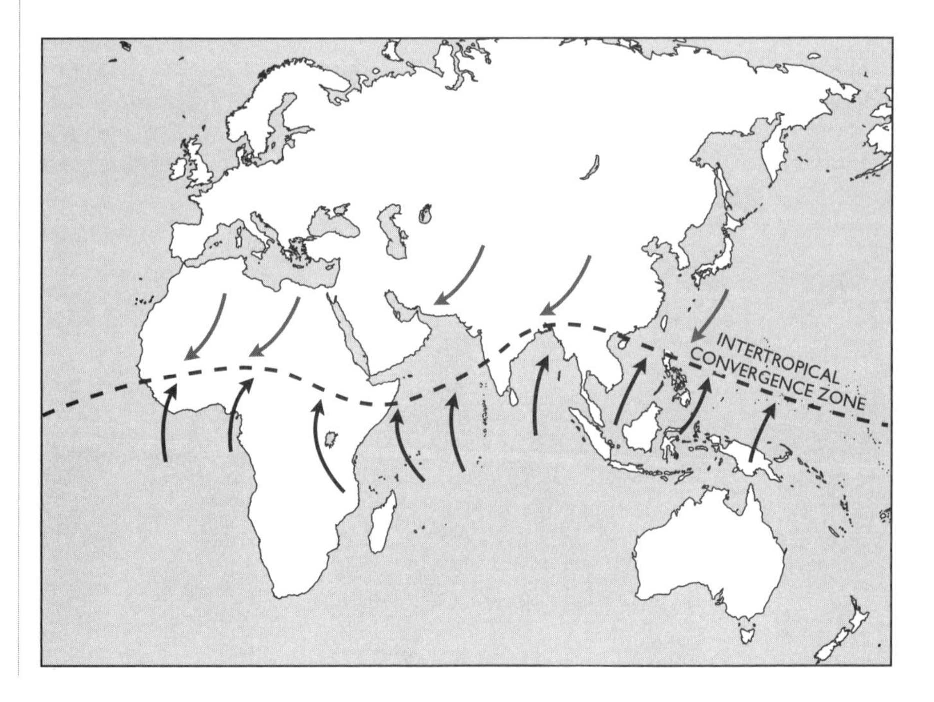

Figure 152 *The monsoons bring life-giving rains to half the people of the world.*

can lead to years of drought or flood, which are normally expected about 30 times a century. If the monsoons fail to arrive and drought conditions strike densely populated regions, millions of people are placed in jeopardy.

Monsoons owe their existence to the temperature difference between land and sea, which causes atmospheric pressure changes that are equalized by the winds. Since Earth's surface is three-quarters water, the oceans absorb great quantities of heat. Water evaporating from the oceans at any given time stores about one-sixth the solar energy reaching the surface of Earth. During a monsoon, part of this enormous heat reservoir is released over the land when water vapor in moist ocean air condenses into rain.

The summer monsoons continue as long as unbalanced forces remain between the land and the ocean. As fall arrives, the ocean temperature drops, reducing the temperature difference between land and sea. The energy of the system then runs down, the monsoon retreats, and the winter dry season begins. With the onset of winter, the land loses heat faster than the ocean. The resulting increased heat loss from the land and the greater heat capacity of the ocean causes the winds to blow in the opposite direction.

The monsoons fail to arrive in Africa because the deflection of the westerlies southward results in a displacement of a high-pressure system hovering over the Sahara Desert. Normally, when the monsoon encounters the southern edge of the high-pressure zone, moisture condenses and rain falls over the Sahel. However, when the high-pressure zone displaces to the south, the monsoon drops its rain before it reaches the Sahel.

The Ghat mountain chain in southwest India plays an important part in forcing moisture-laden air from the Arabian Sea upward, where it cools and releases its rain. The monsoon sweeps northward across India, drenching fields and flooding villages. However, when blocking high-pressure zones lying south of the Himalaya Mountains keep the monsoons away from the Indian subcontinent, droughts sweep through the region. Sometimes the monsoons are delayed for several weeks because the lower temperature of the ocean upwind from the land limits the moisture in the air above the water. When the ocean warms, the monsoons return.

DUST BOWLS

Evidence suggests the American Great Plains experienced tremendous dust storms long before the arrival of pioneer settlers and the introduction of agriculture in the region. Consequently, farming and ranching methods have seriously aggravated the problem. Over the last 150 years, the average soil depth in the most productive areas of the United States has been cut in half by intensive agriculture. Soil erosion is estimated to cost the nation from $30 billion to $45 billion each year in lost productivity. As a result of soil loss from agricultural fields, sediment has been building up in streams, rivers, and floodplains, aggravating their flood potential.

The mid-1930s Dust Bowl was an extended period of drought in the western United States and the nation's worst ecological disaster. Tremendous quantities of topsoil were airlifted out of the Great Plains and deposited downwind, often burying areas under thick layers of sediment (Fig. 153). Massive dust storms raced across the prairie, transporting more than 150,000 tons of sediment per square mile. Since then, improved agricultural practices have reduced this hazard in the United States as well as in other parts of the world. Many regions, however, remain at risk from soil erosion, seriously undermining efforts for populations to feed themselves.

The strong winds of the prairie create gigantic dust storms and severe erosional problems. The tendency of the wind to erode the soil is often aggravated by improper agricultural practices. In the United States, wind erodes about 20 million tons of soil per year. Wind erosion takes out of production an estimated 1.2 million acres of farmland in Russia annually, increasing the difficulty for the nation to feed itself. The primary method of controlling wind erosion is by maintaining a surface cover of vegetation. However, if rainfall is deficient, these measures often fail. The soil simply blows away.

Figure 153 *Farm machinery buried during the 1930s Dust Bowl.*

(Photo courtesy National Center for Atmospheric Research)

Figure 154 *A road adjacent to an unprotected cotton field is buried during a dust storm near Floydada, Texas, on January 25, 1965.*

(Photo by Glen Black, courtesy USDA Soil Conservation Service)

With increasing global temperatures, the central regions of continents that normally experience occasional droughts could become permanently dry wastelands. Soils in most of Europe, Asia, and North America would dry out, requiring additional irrigation. Presently, the United States has more than 500,000 acres of arid and semiarid land. Even larger areas exist in Africa, Australia, and South America. Changes in precipitation patterns would profoundly affect the distribution of the water resources in those areas where they are desperately needed for irrigation of agricultural lands. Rises in temperatures, increases in evaporation rates, and changes in rainfall patterns would also severely limit the export of excess food for developing countries during times of famine.

Subtropical regions might experience a marked decrease in precipitation, encouraging the spread of deserts. Increasing the area of desert and semidesert regions would significantly affect agriculture, causing farmlands to migrate into higher latitudes. Unfortunately, the soils in the northern regions are thin from glacial erosion and would soon wear out from extensive agriculture. In addition, more irrigation would be needed to supply the 1,000 tons of water required to grow a single ton of grain. Furthermore, changing weather patterns due to instabilities in the atmosphere would convert once productive farmlands into deserts (Fig. 154).

MAN-MADE DESERTS

Between 12,000 and 6,000 years ago, many of today's African deserts were lush with vegetation. Grasses and shrubbery covered what is now the Sahara Desert until some unknown environmental catastrophe dried up all the water, leaving behind nothing but sand. A relatively mild arid episode between 7,000 and 6,000 years ago was followed by a severe 400-year-long drought starting 4,000 years ago. Apparently, the monsoon storms that provided water to the Sahara grew weaker, killing off native plants.

The reduction of vegetation further reduced rainfall, producing a vicious cycle of desertification. The drought caused by the vegetation feedback mechanism consequently wiped out almost all plant and animal life in the desert. Such a disaster might have driven entire civilizations out of the desert, forcing them to settle on the banks of the Nile, Tigris, and Euphrates Rivers. North Africa, which is now mostly desert, once had lush grasses and trees in the mountains. It was the breadbasket for Rome, providing grain and meat for the Roman Empire.

The Neolithic, or new stone age, which began around 10,000 years ago following the last ice age, was the beginning of a food-producing revolution. Earth's climate has been unusually mild with few large perturbations, which had a large influence on the rise of civilization. Even in its earliest stages, agriculture was so productive it supported several times more people in a given area than hunting and gathering.

Agriculture had its roots beginning around 15,000 years ago, when primitive peoples stumbled upon the rich Levant region bordering the eastern Mediterranean Sea. This occurred while they were hunting deer and wildebeest and gathering food along the North African coast. The Fertile Crescent, called the cradle of civilization, lies between the Tigris and Euphrates Rivers in present-day Syria and Iraq. The discovery of abundant stands of wild wheat and barley growing in thickets on the uplands was among the most momentous events in human history.

The late Stone Age peoples gathered the wild plants and used primitive stone grinders to process the cereals. The stability of this food supply encouraged people to build permanent settlements. They devised tools to harvest the crop and invented pottery in which to store and cook it. They might have herded gazelle rather than deplete them by overhunting, leading to a new system of animal husbandry. The region became the breadbasket of the Middle East, feeding a population of 17 to 25 million people.

Today, the Fertile Crescent is mostly an infertile desert due to overirrigation and salt accumulation in the soil by Sumerian farmers 6,000 years ago. Heavy use of irrigation, which not long ago turned vast stretches of America's western desert into the world's most productive agricultural land (Fig. 155), is

Figure 155 *Border strip irrigation of crops in Imperial Valley, California.*

(Photo by Robert Brandstead, courtesy USDA Soil Conservation Service)

now ruining hundreds of thousands of acres. Today, humans are repeating the mistakes of the Sumerians.

Around 5,000 years ago, the Phoenicians migrated out of the Sahara Desert and settled along the eastern coast of the Mediterranean Sea. They established such cities as Tyre and Sidon in what is now Lebanon. The land was mountainous and heavily forested with cedars, which became the primary source of timber for the region. When the flat plains along the coast became overpopulated, people moved to the slopes, which they cleared and cultivated, severely eroding the soil. Today, very little remains of the 1,000-square-mile forest. The bare slopes are littered with the remains of ancient terrace walls used in a futile attempt to control erosion.

The remains of several once prosperous cities that are now dead lie in northern Syria. These ancient cities prospered by converting forests into farmland and exporting olive oil and wine. After invasion by the Persians and Arabs, followed by the destruction of agriculture, up to 6 feet of soil eroded from the slopes. Today, after 1,300 years of neglect, the once productive land is nearly completely destroyed, leaving a man-made desert lacking soil, water, and vegetation.

On the plains of Mesopotamia, about 5,000 years ago, nation-states built large irrigation projects that required the hard labor of hundreds of thousands of people and a system of centralized authority to rule over them. What was once a loose-knit egalitarian society was transformed through agriculture into an authoritarian society in a mere 1,000 years, equipped with kings, captains, and slaves. Huge armies of highly organized states fought each other over the control of valuable agricultural land.

Globally, some 7 million square miles or nearly twice the area of the United States has become desertified since the dawn of agriculture, mostly by the abuse of the land. Due to natural processes and human activities, additional land is becoming desertified, amounting to 40 square miles per day or roughly 15,000 square miles a year, about the size of California's Mojave Desert. At the current rate, in the next two or three decades, perhaps as much as 500,000 square miles (about the size of Alaska) of agricultural land will be rendered useless. In North America alone, an estimated 1.1 billion acres have been desertified. Much of the American West 150 years ago was an almost uninterrupted sea of grasslands that have since become desert.

The world's deserts are also getting larger and continue to claim more land. Within the past century, deserts have grown significantly, encroaching upon and eventually consuming neighboring semidesert grasslands. Much of this desertification is due to naturally increasing aridity over the past several thousand years. The North American Desert is the world's fifth largest of its kind. It extends irregularly from east-central Washington to northern Mexico and from the Big Bend of the Rio Grande River in west Texas to the Sierra Nevada of California. It covers some 500,000 square miles, encompassing the Great Basin region and the Sonoran and Mojave Deserts.

The Basin and Range Province of the American Southwest, which includes the Great Basin, contains several mountain ranges. Between these lie dry lake beds called playas that are nearly barren of vegetation. Many basins between ranges are low-lying areas that often contained lakes during wetter climates. Lake-deposited sediments are common, and playas cover the surface. The bodies of water are called alkali lakes because of their high concentrations of salt and other soluble minerals. When the lakes evaporate, they become alkali flats and salt pans such as those in California's Death Valley (Fig. 156).

The process of desertification mainly results from human activity and climate. It severely degrades the environment by removing valuable topsoil, as during the great Dust Bowl years of the 1930s (Fig. 157). Soil erosion removes from production millions of acres of once fertile cropland and pasture every year. Worldwide, perhaps one-third or more of the productive land has been rendered useless by erosion and desertification. Massive dust storms transport the sediment out of the region and deposit elsewhere. After the topsoil erodes, only the coarse sands of the infertile subsoil remain, creating desert conditions.

Tropical rain forests have dwindled over the last few decades. Due to deforestation, deserts have replaced vast expanses of trees. In the Tropics,

Figure 156 *Salt pans and alluvial fans from outwash flows on the floor of Death Valley, Inyo County, California.*

(Photo by H. Drewes, courtesy USGS)

Figure 157 *The 1936 Dust Bowl days in Cimarron County, Oklahoma.*

(Photo by A. Rothstein, courtesy of USDA Soil Conservation Service)

farmers clear much of the land by wasteful slash-and-burn methods. Trees are cut and set ablaze, with their ashes used to fertilize the thin, rocky soil. Extensive agriculture robs the soil of its nutrients. Because most of the world's farmers cannot afford expensive fertilizers, they must abandon their fields and search for new virgin forests to clear.

The denuded land is then subjected to severe soil erosion because of the lack of vegetation to protect against the effects of wind and rain, often leaving bare bedrock behind. Along with the destruction of the rain forests are changing weather patterns within the forests themselves, converting woodlands into deserts. Severe erosion caused by large-scale deforestation clogs rivers with sediment, causing considerable problems for downstream residents. Africa has the worst soil erosion in the world. Its rivers are the most heavily polluted with sediment, whereas other rivers have completely dried out.

Desertification is a global menace. However, it is most severe in central Africa, where the sands of the Sahara Desert, which has greatly expanded over the last two decades, march south across the Sahel region, once a vast stretch of forests and grasslands. Desertification is also self-perpetuating because the light-colored sands reflect sunlight. This creates high-pressure zones that block weather systems from entering the region, which reduces rainfall. The lesser rainfall denudes more land, causing deserts to creep across previously fertile fields.

Deforestation causes the soil to lose much of its capacity to retain moisture, thereby reducing productivity and resistance to drought. African agriculture has never fully recovered from devastating droughts, mainly because of the destruction of the farmland. Primitive farming techniques have devastated the land, causing a decline in African agriculture of about one-quarter since

1960. Under increasing pressure for food production, normally fallow fields are forced into production, which quickly wears out the soil. The root causes of Africa's crisis is growing population, soil erosion, and desertification. Therefore, famine in Africa is becoming more of a man-made disaster.

DUST STORMS

Sandstorms and dust storms (Fig. 158) are awesome meteorological events that play a crucial role in people's physical and economic well-being. They directly threaten life by suffocating people and animals when the air is clogged with large amounts of airborne sediment. Another threat dust storms pose to humans is soil erosion. Every year, additional land becomes desertified. Desertification is also exacerbated by the lack of vegetation, whose roots are needed to hold the soil in place. Furthermore, the land is subjected to flash floods, higher erosion and evaporation rates, and dust storms that transport the soil out of the region.

Human activities account for much of the dust injected into the atmosphere. The concentration of industry in urban areas is a major source of dust. Vehicles inject roadway grit into the air, prompting many municipalities to

Figure 158 *A severe dust storm in Prowers County, Colorado, during the 1930s Dust Bowl.*

(Photo by T. G. Meier, courtesy USDA Soil Conservation Service)

increase street sweeping to reduce the persistent brown haze that hangs over cities. In rural areas, slash-and-burn methods of clearing the land for agriculture expose the bare ground to erosion by the wind, which further clogs the skies with dust.

Dust storms form over deserts during major thunderstorms (Fig 159). They arise in Africa, Arabia, central Asia, central China, and the deserts of Australia and South America, where the most obvious threat is wind erosion. The most immense dust storms result when an enormous airstream moves across the deserts of Africa. Giant dust bands 1,500 miles long and 400 miles wide travel across the region, driven by strong cold fronts.

Some large African storm systems have even carried dust across the Atlantic Ocean to South America, where about 13 million tons land in the Amazon basin annually. The dust over African deserts rises to high altitudes, where westward-flowing air currents transport it across the Atlantic. Fast-moving storm systems in the Amazon rain forest pull in the dust, which contains nutrients that enrich the soil.

Millions of tons of African dust blows across the Atlantic during summer storms and blanket Florida's skies. When the dust settles out, it coats cars and

Figure 159 *The structure of a dust storm.*

other objects with a fine red powder. Other areas along the East Coast of the United States actually violate clean air standards because of the additional load of African dust. Dust from the Sahara Desert blows across the rest of the United States, possibly reaching as far as the Grand Canyon, contributing to the notorious haze that obscures the canyon's beauty. The dust is chemically different from local soils and has a distinctive red-brown color. When added to other air pollutants, the Sahara dust causes a persistent haze, especially in summer.

The African dust has an unexpected benefit, however. The periodic influxes of calcium-rich sediment help regions plagued with acid rain produced from burning fossil fuels by diluting the acidic continent of rainwater. The dust supplies the ocean with much of its iron, an important nutrient needed to keep the marine ecosystem healthy. Coral off the Florida Keys trap the dust inside growth bands, which can be used to trace dust from sources such as storms of sand blowing off the Sahara Desert toward the United States.

Dust storms arise frequently in the Sudan of northern Africa. Near Khartoum, they are experienced about two dozen times a year. They are associated with the rainy season and remove a remarkable amount of sediment. A typical dust storm 300 to 400 miles in diameter can airlift more than 100 million tons of sediment, sufficient to form a pile of dust 2 miles in diameter and 100 feet high. During the height of the season, between May and October, from 12 to 15 feet of sand can pile up against any obstruction exposed to the full fury of the storms.

Severe dust storms also occur in the American Southwest (Fig. 160). Phoenix, Arizona, experiences on average about a dozen per year. As in Africa, American dust storms occur most frequently during the rainy season, normally in July and August. Surges of moist tropical air from the Pacific rush up from the Gulf of California into Arizona and generate long, arching squall lines, with dust storms fanning out in front. These individual outflows often merge to form a solid wall of sand and dust, stretching hundreds of miles. Dust storms also give rise to small, short-lived, and intense whirlwinds within the storms themselves or a short distance out in front called dust devils that can damage buildings and other structures in their paths.

The sediment rises 8,000 to 14,000 feet above ground level and travels at an average speed of 30 miles per hour, with gusts of 60 miles per hour or more possible. The average visibility falls to about one-quarter mile, dropping to zero in very intense storms. After the storm blows away, the skies began to clear in about an hour or so, and visibility returns to normal. If the parent thunderstorm arrives behind the dust storm, its precipitation clears the air more quickly. Often, however, the trailing thunderstorm fails to arrive or the precipitation evaporates before reaching the ground, a phenomenon known as virga. As a result, the sediment remains suspended for several hours or even days.

Figure 160 *A 1935 dust storm in Baca County, Colorado, where it was as dark as midnight for more than an hour.*

(Photo by K. Welch, courtesy USDA Soil Conservation Service)

In dry regions where dust storms were prevalent, the wind transported large quantities of loose sediment. These wind-deposited sand layers are called eolian deposits. Most windblown sediments accumulated into thick deposits of loess (Fig. 161 and 162). This is a fine-grained, loosely consolidated, sheet-

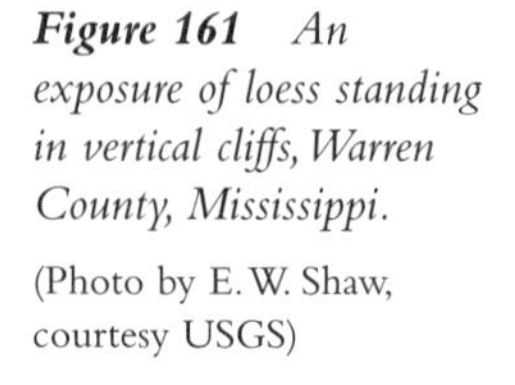

Figure 161 *An exposure of loess standing in vertical cliffs, Warren County, Mississippi.*

(Photo by E. W. Shaw, courtesy USGS)

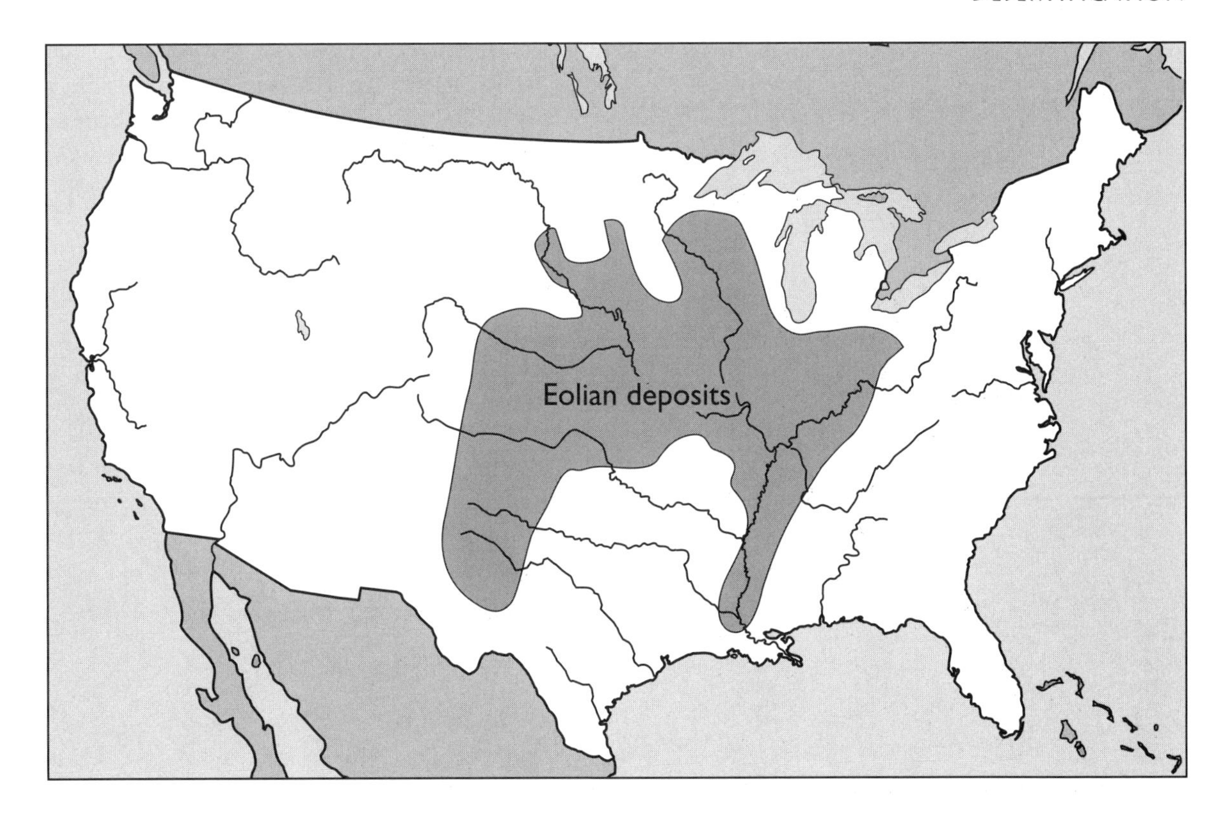

Figure 162 *Windblown soil deposits in the United States.*

like formation that often shows thin, uniform bedding on outcrop. Secondary loess deposits were transported and reworked over a short distance by water or intensely weathered in place.

Loess deposits cover thousands of square miles and were laid down during the ice ages, when continental glaciers swept out of the Arctic regions and buried much of the northern lands. The loess was derived primarily from outwash near major streams that carried glacial meltwater from the front of the glacier. The retreating ice left large, unvegetated areas next to rivers susceptible to wind erosion. As a result, loess deposits rapidly thin with distance from major rivers.

The sediment comprises angular particles of equal grain size composed of quartz, feldspar, hornblende, mica, and bits of clay. It is usually a buff to yellowish brown loamy deposit that is commonly unstratified due to a rather uniform grain size, generally in the silt size range. Loess often contains the remains of grass roots. As with mud bricks, deposits can stand in nearly vertical walls despite their weak cohesion. Loess can also cause problems in construction unless properly compacted because on wetting, it tends to settle.

Loess sediments commonly occur in North America, Europe, and Asia. China contains the world's largest deposits, which originated from the Gobi

Desert and attain hundreds of feet in thickness. Most loess deposits in the central United States are located next to the Mississippi River Valley, where nearly 250,000 square miles is covered by sediment from the glaciated northlands. Deposits also cover portions of the Pacific Northwest and Idaho. Loess makes a yellowish fertile soil responsible for much of the abundant agricultural production of the American Midwest.

SAND DUNES

About 10 percent of the world's arid lands are covered by sand dunes (Fig. 163), driven across the desert by powerful wind currents. Sometimes, sand dunes trample over human settlements and other constructions, often causing considerable damage. The dunes move across the desert floor in response to the wind as sand grains in motion dislodge one another and become airborne for a moment. The size and shape of sand dunes are determined by the direction, strength, and variability of the wind, the soil moisture content, the veg-

Figure 163 *Large dunes in Death Valley, California.*

(Photo courtesy National Park Service)

etative cover, the underlying topography, and the quantity of movable soil exposed to the wind.

As sand dunes march across the desert floor, they engulf everything in their paths. This causes major problems in the construction and maintenance of highways and railroads that cross sandy areas of deserts. Sand dune migration near desert oases creates another serious problem, especially when encroaching on villages. Damages to structures from sand dunes can be reduced by building windbreaks and by funneling sand out of the way. Without such measures, disruption of roads, airports, agricultural settlements, and towns could pose many difficulties for desert regions.

The size and form of sand dunes are determined by the direction, strength, and variability of the wind, the moisture content of the soil, the vegetation cover, the underlying topography, and the amount of movable soil exposed to the wind. Sand dunes generally acquire three basic shapes determined by the topography of the land and patterns of wind flow. Linear dunes (Fig. 146) are aligned in roughly the direction of strong, steady, prevailing winds. They are significantly longer than they are wide and parallel each other, sometimes producing a wavy pattern.

When the wind blows over the dunes' peaks, part of the air flow shears off and turns sideways. The air current scoops up sand and deposits it along the length of the dune, which maintains and lengthens it. The surface area covered by dunes is about equal to the area between dunes. Both sides of the dune are likely steep enough to cause avalanches. The sliding sand grains often produce an unexplained phenomenon known as booming sands. At least 30 booming dunes have been found in deserts and on beaches in Africa, Asia, North America, and elsewhere.

The sound occurs almost exclusively in large, isolated dunes deep in the desert or on back beaches well inland from the coast. The dunes can sound like bells, trumpets, pipe organs, foghorns, cannon fire, thunder, buzzing telephone wires, or low-flying aircraft. The grains in sound-producing sand are usually spherical, well rounded, and well sorted, or of equal size. The booms can be triggered by simply walking along the dune ridges. The low-frequency sound appears to originate from a cyclic event occurring at an equally low frequency. However, normal landsliding involves a mass of randomly moving sand grains that collide with a frequency much too high to produce such a peculiar booming noise.

Crescent dunes, also called barchans, are symmetrically shaped with horns pointing downwind. They travel across the desert at speeds of up to 50 feet a year. Parabolic dunes form in areas where sparse vegetation anchors the side arms, while the center blows outward and moves sand in the middle forward. Star or radial dunes form by shifting winds that pile sand into central points that can rise 1,500 feet and more, with several arms projecting outward, resembling giant pinwheels. Sand also accumulates in flat sheets or forms stringers downwind that do not exhibit any appreciable relief in sand seas.

After discussing the role desertification plays in people's lives, the next chapter deals with the depletion of natural resources and explores other forms of energy.

9

NATURAL RESOURCES

DEPLETION OF INDUSTRIAL MATERIALS

This chapter examines Earth's finite and renewable resources, conservation, and new forms of energy. A fundamental concept of environmental geology implies that natural resources are limited. Resources such as oil, gas, and minerals are recycled so slowly in the geologic cycle they are essentially nonrenewable and therefore finite. By definition, reserves are known and identified earth materials for immediate extraction and use, whereas resources are reserves that can be later extracted. Because resources are indeed limited, important questions arise about their long-term use. Without reliable reserves at home, the United States could become dangerously dependent on foreign sources for meeting its demand.

The rise in industrialization heavily depended on rich stores of natural resources. The exploitation of minerals and energy has substantially improved people's lives. Unfortunately, the depletion of natural resources could threaten future advancement. Many high-grade ore deposits have been heavily exploited and could be mined out in the foreseeable future. The consumption of mineral ores to maintain a high standard of living in the industrialized world and to improve the quality of life in developing countries could hasten the depletion of known economic ore reserves. Then low-grade deposits would have to be worked, dramatically increasing the cost of living.

ENERGY

Worldwide energy consumption has grown exponentially since the Industrial Revolution. It could rise by more than 50 percent in the early part of this century, when petroleum supplies might fail to meet the demand from growing industrial activities. The consumption of petroleum in the industrialized nations is expected to increase significantly. Furthermore, developing countries desire industrialization to improve their standards of living, which has the added benefit of reducing population growth.

Most countries realize that energy development is fundamental to the development of other natural resources and represents an asset that should be properly regulated and sustained. Available natural resources must be understood and managed appropriately if developing countries are to become self-sustaining. Several nations have focused their developmental strategies on specific applications such as those directly related to energy sources.

Almost half of the world's primary energy is supplied by petroleum, which includes oil and natural gas (Fig. 164). Presently, more than 1 trillion barrels of oil have been discovered, of which fully one-third or more has been

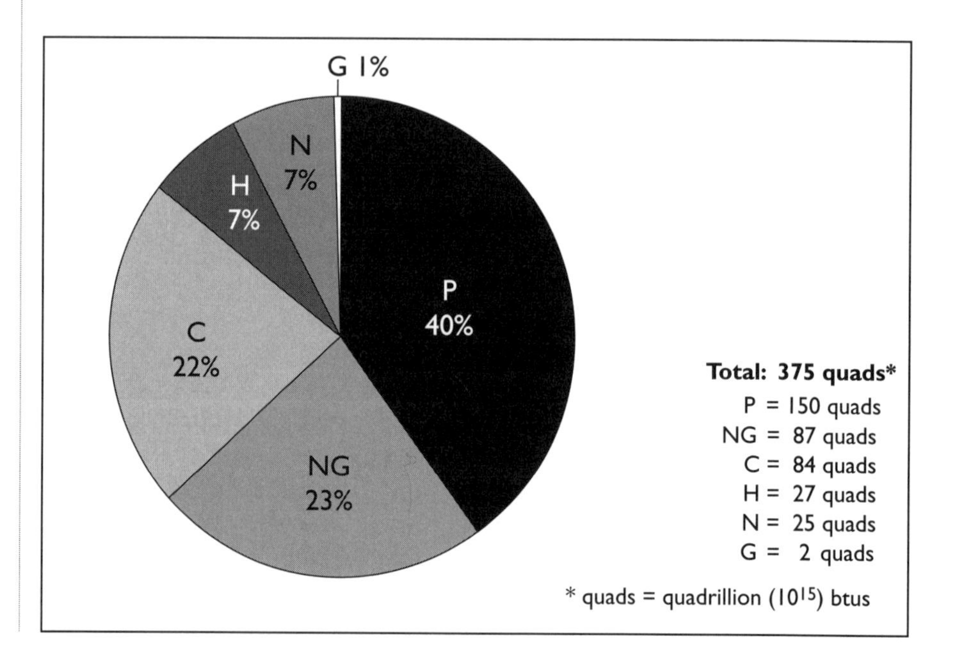

Figure 164 *Sources of primary energy produced internationally—petroleum (P), natural gas (NG), coal (C), nuclear (N), hydroelectric (H), and geothermal, solar, wind, and wood and waste electric power (G).*

depleted. The world consumes about 70 million barrels of oil daily, with the United States using nearly one-third of the total. An average American consumes more than 40 barrels of oil a year compared with the average European or Japanese who uses between 10 and 30 barrels annually. In contrast, an average person in a developing country uses the equivalent of only 1 or 2 barrels of oil yearly.

The creation of reservoirs of oil and natural gas requires a special set of circumstances, including a sedimentary source for the oil, a porous rock to serve as a reservoir, and a confining structure to trap the oil. The source material is organic carbon trapped in fine-grained, carbon-rich sediments. Porous and permeable sedimentary rock such as sandstones and limestones form the reservoir. Geologic structures produced by folding or faulting of sedimentary beds trap or pool the oil. Petroleum often associates with thick beds of salt. Because salt is lighter than the overlying sediments, it rises toward the surface, creating salt domes that help trap oil and natural gas.

The organic material originates from microscopic organisms living primarily in the surface waters of the ocean and concentrates in fine particulate matter on the ocean floor. The transformation of organic material into oil requires a high rate of accumulation or a low oxygen content in the bottom water to prevent oxidation of organic material before burial under layers of sediment. Oxidation causes decay, which destroys organic matter. Therefore, areas with high rates of accumulation of sediments rich in organic material are the most favorable sites for the formation of oil-bearing rock. Deep burial in a sedimentary basin heats the organic material under high temperatures and pressures, which chemically alters it. Essentially, the organic material is cracked into hydrocarbons by the heat generated in Earth's interior. If the hydrocarbons are overcooked, natural gas results.

The hydrocarbon volatiles along with seawater locked up in the sediments migrate upward through permeable rock layers. They accumulate in traps formed by sedimentary structures that provide a barrier to further migration. Without such a cap rock, the volatiles continue rising to the surface and escape into the ocean from natural seeps, amounting to about 1.5 million barrels of oil yearly. This is minuscule compared with some 25 million barrels of oil accidentally spilled into the sea (Fig. 165). From several tens of millions to a few hundred million years are needed to produce petroleum, which mainly depends on the temperature and pressure conditions within the sedimentary basin.

During offshore exploration, the geology of the ocean floor is determined to test whether the proper conditions exist for trapping oil and gas. This testing greatly aids oil companies in their exploration activities. Petroleum exploration begins with a search for sedimentary structures conducive to the formation of oil traps. Seismic surveys delineate these structures by using

Figure 165 *Workers clean the beaches of Sandy Hook, New Jersey, from a major oil spill in March 1980.*

(Photo courtesy U.S. Coast Guard)

explosions that generate waves similar to sound waves that are received by hydrophones towed behind a ship. The seismic waves reflect and refract off various sedimentary layers, providing a sort of geologic CAT scan of the ocean crust.

After choosing a suitable site, the oil company sets up an offshore drill rig (Fig. 166). This stands on the ocean floor in shallow water or free floats anchored to the bottom in deep water. While drilling through the bottom sediments, workers line the well with steel casing to prevent cave-ins and to act as a conduit for the oil. A blowout preventer placed on top of the casing prevents the oil from gushing out under tremendous pressure once the drill bit penetrates the cap rock. If the oil well is successful, additional wells are drilled to develop the field.

Over the last several decades, offshore drilling for oil and natural gas in shallow coastal waters has become extremely profitable. About 20 percent of the world's oil and 5 percent of the natural gas production is offshore. In the future, twice as much oil might be pumped from offshore than from land. Unfortunately, as much as 2 million tons of offshore oil spills into the ocean each year. Oil spills of this magnitude could create an enormous environmental problem as production rises.

Offshore oil drilling began in the mid-1960s and escalated a decade later following the 1973 Arab oil embargo. This created shortages, which tripled the price of crude oil and caused American motorists to stand in long lines at gas stations. Important finds such as Prudhoe Bay on Alaska's North Slope (Fig. 167) and the North Sea off Great Britain resulted from intensive exploration for new reserves of offshore petroleum. The desire for energy independence prompted oil companies to explore for petroleum in the deep oceans. Many difficulties were encountered, however. These included storms at sea and the loss of personnel and equipment, which could not justify the few new discoveries made. Futuristic plans are to build drilling equipment and workrooms on the seafloor where they are not affected by storms, making some deep-sea oil and gas fields available for the first time.

Oil production in the upcoming years could eventually level off and begin to decline. As a result, alternative fuels would have to be developed to meet the continuing demand for energy. Oil-importing countries, which

Figure 166 *A semisubmersible drilling rig in the Mid-Atlantic outer continental shelf.*
(Photo courtesy USGS)

Figure 167 *Oil well drilling on Alaska's North Slope, Barrow district, Alaska.*

(Photo by J. C. Reed, courtesy U.S. Navy and USGS)

consume about half the petroleum on the market, would require a transition from a dependence on oil to a greater reliance on other fossil fuels, nuclear energy, and renewable energy sources.

A hybrid of natural gas and geothermal energy are reservoirs of hot, gas-charged seawater called geopressured deposits (Fig. 168) lying beneath the Gulf Coast off Texas and Louisiana. The gas deposits formed millions of years ago when seawater permeated porous beds of sandstone between impermeable clay layers. The seawater captured heat building up from below and dissolved methane from decaying organic matter. As more sediments piled on, the hot, gas-charged seawater became highly pressurized. Wells drilled into this formation tap high-temperature steam along with natural gas. The gas could provide an energy potential equal to about one-third of all coal deposits in the United States, bringing the nation closer to energy self-sufficiency.

Another potential source of energy is a snowlike natural gas called methane hydrate on the deep-ocean floor. Methane hydrate is a solid mass formed when high pressures and low temperatures squeeze water molecules into a crystalline cage around a methane molecule. Vast deposits of methane hydrate are thought to be buried in the seabed around the continents and represent the largest untapped source of fossil fuel left on Earth. Methane hydrate hidden beneath the waters around the United States alone hold enough natural gas to supply all the nation's energy needs for 1,000 years.

Figure 168 *A drilling rig used to extract geopressured energy near Houston, Texas.*

(Photo courtesy U.S. Department of Energy)

Tapping into this enormous energy storehouse, however, is costly and potentially dangerous. If the methane hydrate becomes unstable, it could erupt like a volcano. Several craters on the ocean floor are identified as having been caused by gas blowouts. Giant plumes of methane have been observed rising from the seabed. Methane escaping from the hydrate layer also nourishes microbes that, in turn, sustain cold-vent creatures such as tube worms. Additionally, methane, a potent greenhouse gas, escaping into the atmosphere could escalate global warming.

At the present rate of consumption, the easily tapped petroleum reserves could be depleted by the middle of this century. Unless safe alternatives such as fission, fusion, solar, and geothermal energy are developed and rapidly exploited, industrial plants might have to convert to coal. The resources of coal are substantial. However, the environmental impacts of burning coal are much more severe than oil or natural gas.

About one-quarter of the world's energy is supplied by coal. Peak usage began during the 1920s, when coal accounted for more than 70 percent of all fuel consumption and most air pollution. The United States has increased its consumption by about 70 percent since the 1970s, mostly for coal-fired electrical generation. Electrical generating plants account for about 75 percent of the coal consumed in the nation. Coal-fired power plants provide about half America's electrical energy production. However, the consumption of coal is bound to rise as the cost of natural gas continues to increase. Utility companies favor natural gas because it is much cleaner burning than coal.

The total world coal production is about 5 billion tons annually, with the United States accounting for about half the coal mined and consumed by the free world. To keep up with an increasing demand, the United States would have to mine 50 percent more coal than is mined at present. Most of the world's coal deposits are barely touched. Coal reserves far exceed all other fossil fuels combined and are sufficient to support a large increase in consumption well into this century. Abundant coal reserves exist in the western United States, Canada, Soviet Union, Asia, and South Africa. The economically recoverable reserves worldwide are estimated at nearly 1 trillion tons. At the present rate of consumption, resources could possibly last another two centuries.

The United States possesses nearly half the world's economic coal reserves (Fig. 169). Because coal is the cheapest and most abundant energy source, it will be the most favorable alternative fuel to replace costly petroleum when supplies run low. The conversion to coal could, however, create a tremendous overloading of carbon dioxide and other dangerous chemicals in the atmosphere. Coal combustion yields twice as much carbon dioxide per unit of energy than oil and natural gas. Therefore, a dramatic switch to coal could contribute substantially to the greenhouse effect and influence the global climate.

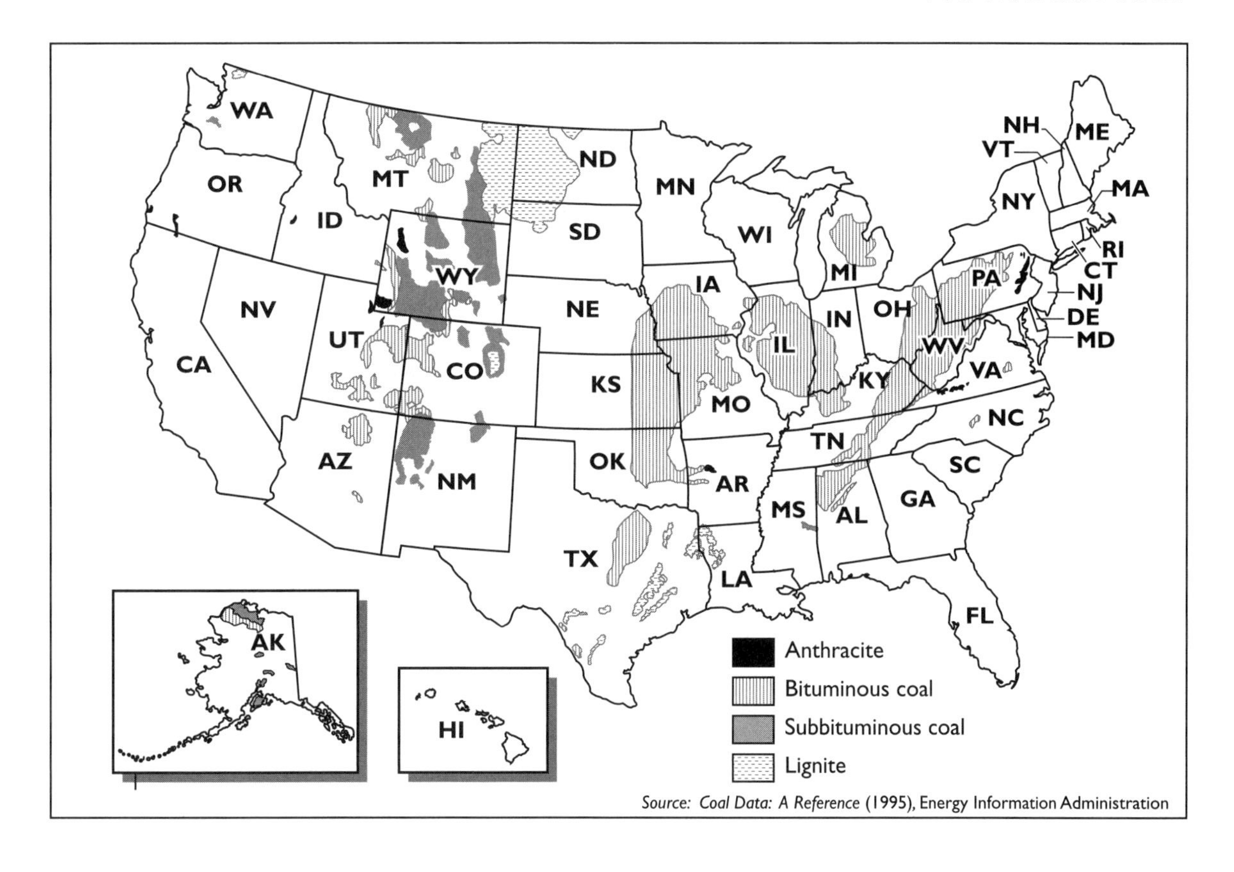

Figure 169 *Coal deposits in the United States.*

The long-term increase in atmospheric carbon dioxide, as much as 25 percent since 1860, is the result of an accelerated release of carbon dioxide by fossil fuel combustion. For every ton of carbon in fossil fuels, more than 3.5 tons of carbon dioxide is liberated during combustion. The combustion of coal also produces sulfur dioxide and nitrogen oxides discharged into the atmosphere, which contribute to acid precipitation. The coal from eastern underground mines (Fig. 170) generally has a high sulfur content, requiring coal-fired power plants to install smokestack scrubbers to reduce emissions of sulfur products that cause acid rain. Alternately, utilities could burn low-sulfur coal extracted from massive open-pit mines in the West (Fig. 171), which is much more expensive.

Huge untapped reserves of oil exist in oil-shale deposits in the western United States (Fig. 172). Their potential oil content exceeds that of all other petroleum resources in the entire world. Large tar sand deposits, such as those in California and Alberta, Canada, are another potential source of petroleum equal to about 500 billion barrels of oil, once extraction becomes economical.

Figure 170 *Underground coal mining near Benton, Illinois.*

(Photo courtesy U.S. Department of Energy)

Figure 171 *Open-pit coal mining at the Absaloka mine, Montana.*

(Photo courtesy USGS)

Figure 172 *An outcrop of oil shale, Uintah County, Utah.*

(Photo by D. E. Winchester, courtesy USGS)

MINERALS

This world is rich in natural resources. The exploitation of minerals and energy has greatly improved people's standards of living. Unfortunately, the depletion of natural resources could threaten future advancement. Many high-grade ore deposits have been heavily exploited and might soon be mined out. The consumption of mineral ores to maintain a high standard of living in the industrialized world and to improve the quality of life in developing countries might lead to the depletion of known high-grade ore reserves by the middle of the century (Table 14). Then low-grade deposits would have to be worked, dramatically increasing the cost of goods and commodities.

Fortunately, humans have barely scratched the surface in the search for mineral deposits. Improved techniques in geophysics, geochemistry, and mineral exploration has helped keep resource supplies up with rising demand. As improved exploration techniques become available, future supplies of minerals will be found in yet unexplored regions. Immense mineral resources lie at great depths, awaiting the mining technology to recover them.

Joining the geologist in this search for new energy and mineral deposits are several types of remote sensors on aircraft and satellites. Mineral deposits reveal themselves in many different ways. Most are invisible to the naked eye but are detectable in various sensors operating at wavelengths outside the visible spectrum. Satellite imagery can delineate geologic structures such as faults, fracture zones, and contacts in which mineral ores are deposited and appear on satellite imagery as distinctive lineaments. Other structures, such as folds or domes, are also distinguishable in satellite imagery and might serve as traps for oil and gas.

TABLE 14 NATURAL RESOURCE LEVELS (DEPLETION RATE IN YEARS AT PRESENT CONSUMPTION)

Commodity	Reserves*	Total resources
Aluminum	250	800
Coal	200	3,000
Platinum	225	400
Cobalt	100	400
Molybdenum	65	250
Nickel	65	160
Copper	40	270
Petroleum	35	80

*Reserves are recoverable resources with today's mining technology.

Mineral deposits might also be detected as discolorations of the surrounding rock or by particular types of vegetative growth, which reflect certain soil types. The soils, in turn, are determined by their mineral content derived from the underlying parent rock. Radar can penetrate heavy cloud cover and the vegetation canopy to observe the ground. The radar data is particularly useful for identifying structures and classifying rock units. Precision radar altimetry from satellites and other remote sensing techniques can map the ocean bottom, where a large potential for the world's future supply of minerals and energy exists.

Minerals are homogenous substances, with unique chemical compositions and crystal structures. Most minerals develop crystals, which greatly aid in their identification. The most abundant rock-forming minerals are quartz and feldspar. These make up most of the noncarbonate, or crystalline, rocks. When a magma body cools, a variety of minerals with varying crystal sizes separate out of the melt. This leaves behind highly volatile mineralized fluids that invade the country (host) rocks surrounding the magma chamber to form veins of ore, from which the mineral can be extracted. Single-element minerals can form metallic ores such as copper or nonmetallic substances such as sulfur, which is mostly associated with volcanic activity.

Ores are naturally occurring materials from which valuable minerals are extracted. Mineral ore deposits form very slowly, taking millions of years to create an ore significantly rich to be suitable for mining. Certain minerals precipitate over a wide range of temperatures and pressures. They commonly occur with one or two minerals predominating in sufficiently high concentrations to make their mining profitable.

Extensive mountain building activity, volcanism, and granitic intrusions provide vein deposits of metallic ores. Copper, tin, lead, and zinc ores concentrate directly by magmatic activity, when magma bodies invade Earth's crust. These concentrations form as hydrothermal (hot-water emplaced) vein deposits, which are mineral fillings precipitated from hot waters percolating along underground fractures.

Hydrothermal ore deposits are a major source of industrial minerals. They are so valuable that an intense study of their genesis has been conducted for more than a century. Their origin is by precipitation from solutions within the upper few miles of Earth's crust. Ore deposition occurs from rising hot solutions heated by active magmatism, with temperatures exceeding 600 degrees Celsius. Many metals occur in ores as sulfides or oxides that are generally insoluble except under certain physical and chemical conditions. Without ample sulfides, metals join with ordinary rock-forming minerals in trace concentrations that are much too low to be recovered economically. The ore precipitates in fractured rock near the surface, where the hydrothermal fluids flow along restricted channels.

Toward the turn of the 20th century, geologists found that hot springs at Sulfur Bank, California, and Steamboat Springs, Nevada (Fig. 173), deposited

Figure 173 *Steam fumaroles at Steamboat Springs, Nevada.*

(Photo courtesy USGS)

the same metal-sulfide compounds found in ore veins. Therefore, if the hot springs were depositing ore minerals at the surface, hot water must be filling fractures in the rock with ore as it moves toward the surface. The American mining geologist Waldemar Lindgren discovered rocks with the texture and mineralogy of typical ore veins by excavating the ground a few hundred yards from Steamboat Springs. He proved that many ore veins formed by circulating hot water called hydrothermal fluids. The mineral fillings precipitated directly from hot waters percolating along underground fractures.

The rocks surrounding the magma chamber are possibly the true source of the minerals found in hydrothermal veins. In this case, the volcanic rocks act only as a heat source that pumps groundwater into a giant circulating system. Cold, heavier water moves down and into the cooling volcanic rocks carrying trace amounts of valuable elements leached from the surrounding rocks. When heated by the magma body, the water becomes more buoyant and rises into the fractured rocks above. After cooling and losing pressure, the water precipitates its mineral content into veins and moves down again to pick up another load of minerals.

A gigantic subterranean still is supplied with heat and volatiles from magma chambers. As the magma cools, silicate minerals such as quartz crystallize first, leaving behind a concentration of other elements in a residual melt. Further cooling of the magma causes the rocks to shrink and crack. This allows the residual magmatic fluids to escape toward the surface and invade the surrounding rocks, forming veins. Certain minerals precipitate over a wide range of temperatures and pressures. They commonly occur with one or two minerals predominating in sufficiently high concentrations to make their mining profitable.

A second type of mineral ore emplacement is called massive sulfide deposits. These deposits originated on the ocean floor at midocean spreading centers and occur as disseminated inclusions or veins in ophiolite complexes (Fig. 174) exposed on dry land during continental collisions. One of the most noted deposits is in the 100-million-year-old Apennine ophiolites, which were first mined by the ancient Romans. Massive sulfide deposits are mined extensively in other parts of the world for their rich ores of copper, lead, zinc, chromium, nickel, and platinum.

ORE DEPOSITS

Iron is one of the most important ore deposits and largely responsible for the Industrial Revolution. Economic deposits of iron ore are found on all continents. With today's technology, ore grades must generally exceed 30 percent to make mining profitable. Layered deposits of iron oxide cover

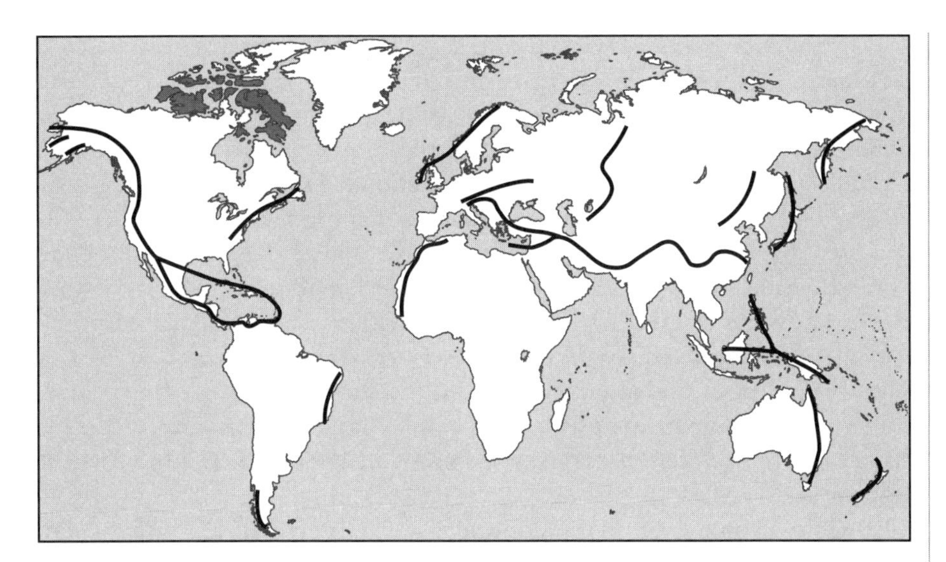

Figure 174 Worldwide distribution of ophiolites, which are slices of oceanic crust shoved up onto land by plate tectonics.

huge regions, such as the Lake Superior region of North America and the Hamersley Range of western Australia. The Mesabi Range of northeast Minnesota is the major supplier of iron ore for the United States. The ore occurs in a banded-iron formation laid down more than 2 billion years ago. The Clinton iron formation is the chief ore producer for the Appalachian region. The iron occurs in an oolithic ironstone deposited more than 400 million years ago.

Although basalt comprises about 5 percent iron, oxidized lavas are not usually mined for their iron content. Basaltic lavas are oxidized when steam or other gases pass through the rock when it is still in a highly fluid state, producing iron minerals. The El Laco mine on the border between Chile and Argentina is a rarity among iron mines. The ore body is a large lava flow consisting almost entirely of the iron minerals hematite and magnetite. The iron was concentrated in a homogenous fluid saturated with water vapor that erupted onto the surface as an iron lava.

Zambia's great copper belt is estimated to contain one-quarter of the world's copper. The Keweenaw peninsula in the Lake Superior region comprises a 100-mile-long and 3-mile-wide copper belt some 2 billion years old. As many as 400 individual basalt flows totaling about 20,000 feet thick contain copper concentrated at the tops of each lava flow. Hydrothermal fluids of copper sulfide arose from instrusive rocks underlying the basalts and were emplaced into the interventing layers of lava. Oxygen derived from iron oxides within the basalt combined with sulfur from the copper minerals to reduce them to metallic copper chemically.

Rich lead and zinc hydrothermal deposits exist in the Tri-State Region of the Mississippi River Valley near Missouri. Copper, tin, lead, and zinc ores concentrate directly by magmatic activity, forming hydrothermal vein deposits. Economic deposits of gold, silver, lead, zinc, and copper exist in the Cordilleran mountain regions of North and South America.

On opposite extremes of the hydrothermal spectrum are mercury and tungsten. Mercury is a liquid at room temperature. Tungsten is one of the hardest metals, which makes it valuable for hardening steel. All belts of productive deposits of mercury are associated with volcanic systems. Mercury forms a gas at low temperatures and pressures. Therefore much of Earth's mercury is lost at the surface from volcanic steam vents and hot springs. Tungsten, by comparison, precipitates at very high temperatures and pressures often at the contact between a chilling magma body and the rocks it invades.

Gold is mined on every continent except Antarctica. In Africa, the best gold deposits are in rocks as old as 3.4 billion years. In North America, the most productive gold mines are in the Great Slave region of northwest Canada, where more than 1,000 deposits are known. These gold deposits are found in greenstone belts invaded by hot magmatic solutions from the intrusion of granitic bodies. The gold occurs in veins associated with quartz. In Chile, silver and gold were mined from the eroded stumps of ancient volcanoes. Cerro Rico, whose name means "rich hill," in Bolivia is a 15,000-foot volcano literally shot through with veins of rich silver one, some more than 12 feet thick. A variety of other metallic deposits lies in the mountains of southern Europe and in the mountain ranges of southern Asia as well. The world's largest nickel deposit at Sudbury, Canada, is thought to have been created by a massive meteorite impact about 1.8 billion years ago.

Half the world's production of chromium comes from South Africa, which is also responsible for much of the global diamond production. The diamonds are disseminated in a volcanic structure called a kimberlite pipe that resembles a funnel reaching deep into Earth's mantle. Most kimberlite pipes are about 100 million years old, although the diamonds they hold formed billions of years ago under great temperatures and pressures. The major platinum deposits of the world include the Bushveld Complex of South Africa and the Stillwater Complex of Montana.

One of the most important industrial nonmetallic minerals is sulfur. Because sulfur occurs in abundance in other geologic settings, volcanoes contribute only a small proportion of the world's economic requirements. The largest volcanic sulfur mines are in northern Chile. The open-pit mine atop the Aucanquilcha Volcano has the distinction of being the highest mine in the world, lying some 20,000 feet in elevation. The mine lies within the core of a complex andesite volcano. The entire central part contains ore with a 30 percent sulfur content.

Valuable deposits of phosphate used for fertilizers are mined in Idaho and adjacent states. Evaporate deposits in the interiors of continents, such as the potassium deposits near Carlsbad, New Mexico (Fig. 175), indicate these areas were once inundated by ancient seas. Thick beds of gypsum used in the manufacture of plaster of Paris and drywall board were also deposited in the continental interiors. Minerals such as sand and gravel, clay, salt, and limestone are mined in large quantities throughout the world.

The most promising mineral deposits are manganese nodules on the ocean floor (Fig. 176). They are particularly well developed in deep, quiet waters far from continental margins and active volcanic zones. Concentric layers accumulate over millions of years until the nodules reach about the size of a potato, giving the ocean floor a cobblestone appearance. A ton of manganese nodules contains about 600 pounds of manganese, 29 pounds of nickel, 26 pounds of copper, and 7 pounds of cobalt. However, their location at depths approaching 4 miles makes extraction on a large scale extremely difficult.

Figure 175 *The Duval Sulphur and Potash Company's mining operation near Carlsbad, New Mexico.*

(Photo by E. F. Patterson, courtesy USGS)

Figure 176 *Manganese nodules on Sylvania Guyot, Marshall Islands, at a depth of 4,300 feet.*

(Photo by K. O. Emery, courtesy USGS)

CONSERVATION

The human race is on a collision course between limited resources and the growing numbers of people using them. A population exceeds carrying capacity—the ability of the land to provide for people's needs—when it cannot be maintained without rapidly depleting nonrenewable resources and degrading the environment. As world populations continue growing geometrically on a planet whose resources are dwindling rapidly, the vast majority of people are forced to live barely a subsistence level of life.

The western world is rapidly devouring natural resources at a high rate. One-fifth of the human population lives in the relatively few rich nations of the Northern Hemisphere. In contrast, most people inhabit poverty-stricken countries, mainly in the Southern Hemisphere, often referred to as the "poor south." The rich nations consume about 80 percent of Earth's natural resources and are directly or indirectly responsible for most of the pollution and degradation of the environment.

An increase in the efficiency of energy use and the utilization of alternative fuels would help improve the environments of the developed nations. Such an effort would also help developing countries raise their standards of living without a significant increase in energy use and a corresponding rise in pollution. Failure to take these measures could condemn 80 percent of the human population to substandard living conditions.

Industrial energy consumption per unit of production and the per capita consumption of resources and production of pollution is about four times greater in the United States than in other modern, industrialized nations. A marked improvement in energy efficiency can cut industrial air pollution by upward of 50 percent. By improving insulation and using more efficient construction materials, appliances, and lighting in buildings and homes, energy consumption and air pollution in those facilities can be cut in half.

The buildings sector of the economy is the single largest consumer of energy in the United States, comprising 40 percent of the total energy budget. Buildings consume three-quarters of all electricity generated in this country. Over a building's life span, the energy bill can exceed twice the construction cost. Meanwhile, the transportation sector consumes about 200 billion gallons of fuel each year and produces over half the air pollution generated by fossil fuel combustion. An improvement of 5 miles per gallon in American automobile mileage would cut carbon dioxide emissions by nearly 100 million tons a year. Energy-efficient automobiles, including electric cars (Fig. 177), would cut automotive carbon dioxide by up to 70 percent. Car pooling and mass transit would reduce smog in big cities. Furthermore, the use of alternative fuels such as natural gas and methanol would cut emissions while reducing dependence on foreign sources of petroleum.

Coal is the most abundant fossil fuel. However, it is also the dirtiest in terms of emissions of particulate matter and carbon dioxide as well as aerosols composed of oxides of nitrogen and sulfur that produce acid rain. Nevertheless, coal can be burned more efficiently in pressurized, fluidized bed boilers. These burn most of the pollutants, cut nitrogen oxides by one-third, and reduce sulfur emissions by more than 90 percent compared with conventional power plants. Moreover, pollution controls installed on existing coal-fired power plants could cut nitrogen oxides and sulfur dioxides by another 90 percent. Such improvements have significantly reduced acid rain in the United States.

Figure 177 *An electric automobile is tested at the Idaho Laboratory Facility, Idaho Falls, Idaho.*

(Photo courtesy U.S. Department of Energy)

Natural gas composed mostly of methane is the second most plentiful hydrocarbon energy source in the nation. Switching to natural gas where possible would cut carbon dioxide emissions in half. Unfortunately significant amounts of natural gas leak into the atmosphere during transmission and distribution, possibly contributing to greenhouse warming. Electrical generating plants and motor vehicles could take advantage of this fuel. Compressed natural gas mixed with hydrogen yields the cleanest-burning alternative fuel for powering motor vehicles. Present reserves of natural gas can withstand steep increases well into this century. Natural gas could also be supplemented with methane generated by the conversion of waste products (Fig. 178).

Instead of burning fuel in separate plants to generate electricity and manufacture products or heat buildings, efficiency is substantially improved when operations are combined in a process called cogeneration rather than allowing the waste heat simply to escape into the atmosphere. Cogeneration could boost total efficiency by up to 90 percent and cut air pollution in half. These conservation methods could curtail the effects of global warming by improving energy efficiency and developing nonpolluting substitute energy sources. With the conservation of natural resources and the exploration of alternative energy sources, the wealth of the world would be preserved for future generations.

RECYCLING

One solution to the chronic garbage disposal problem is recycling (Fig. 179). It not only reduces trash dumped into landfills but also generates no pollution while decreasing the need to mine or harvest new raw materials, thereby saving the environment. Recycling also reduces the demand for incineration and the pollution problems it entails. A more acceptable solution to the growing disposal problem is recycling along with a concurrent reduction of convenience packaging and redundant products.

To reduce the amount of garbage, the U.S. economy, which is based on overconsumption and waste, would have to be overhauled. Such an economic system might be improved by including higher taxes on packaging, banning certain unrecyclable plastics and throwaway products, and instituting standards for making products last longer. Tax breaks might encourage industries to use recyclable materials. Manufacturers must be discouraged from making durable goods that do not last or waste energy and encouraged to use recycled materials whenever possible. These steps could be taken without requiring major changes in lifestyle while vastly improving the environment.

About 80 percent of municipal solid waste is recyclable material. However, implementing recycling on a national scale is difficult because many industries refuse to use secondary materials. Furthermore, companies need assurances that the supply of recyclable materials is abundant and reliable. Recycling induces little economic incentive when alternative methods of waste disposal, such as incineration, remain attractive. However, incineration

Figure 178 *A facility designed to convert animal wastes into methane gas, Barton, Florida.*

(Photo courtesy U.S. Department of Energy)

Figure 179 *Aluminum beverage cans are emptied into a crushing and baling machine at the Pensacola Naval Air Station, Florida.*

(Photo by Jim Bryant, courtesy U.S. Navy)

could be reduced or avoided entirely by aggressive recycling. The recycling of waste plastics, accounting for about 40 percent of landfill trash, can yield a high-quality fuel oil to relieve petroleum imports.

RENEWABLE ENERGY

Industrial nations face an energy crisis of immense proportions if alternative energy sources are not found and rapidly exploited before supplies of fossil fuels run low. Nuclear energy is one of the best solutions to the world's chronic energy problems. Many European countries, particularly France, rely heavily on nuclear energy to replace costly fossil fuels. To combat atmospheric pollution and global warming, a reassessment of nuclear energy is recommended. Nuclear electrical generating plants are essentially nonpolluting because they do not produce greenhouse gases. The safety of the plants must be ensured to prevent nuclear accident. Nuclear wastes have to be managed properly if nuclear energy is to be considered a viable alternative energy source.

Fusion nuclear energy (Fig. 180) is a renewable resource and essentially nonpolluting. It is safer than fission nuclear energy. Its by-products are energy and helium, a harmless gas that escapes into space. Many advances in fusion research have been made. However, a workable electrical generating station is still far into the future. Unless a major breakthrough occurs soon and the technology is rapidly commercialized, fusion will not figure significantly into the world's energy needs of the near future.

Figure 180 *The Omega 24 system at the University of Rochester, New York, is used to study laser fusion.*

(Photo courtesy U.S. Department of Energy)

Solar energy is another successful energy alternative. The sunlight striking Earth is thousands of times greater than the world's present energy usage. Photovoltaic or solar cells convert sunlight directly into electricity, with an efficiency of about 20 percent. Manufacturing these solar cells is very expensive, making large-scale use uneconomical. However, solar cells with a lower efficiency can be manufactured in mass quantities at greatly reduced prices.

The conversion of sunlight into electricity can also be achieved by vast arrays of solar collectors combined in solar farms (Fig. 181). Sunlight is focused into a powerful narrow beam by banks of heliostatic mirrors that automatically track the Sun as it travels across the sky. The light beam is directed onto a central receiving station, where the intensified light heats a boiler and the superheated steam drives a turbine generator. Presently, solar power stations cannot compete economically with conventional fossil-fuel-generating plants. However, they could become more viable as fossil fuels grow scarce and expensive.

Buildings can use solar energy to supplement conventional furnaces and water heaters (Fig. 182). This would provide substantial savings on utility bills while conserving nonrenewable energy resources. The Sunbelt states, blessed with a generous supply of sunlight, can take full advantage of this form of solar energy. The systems usually pay for themselves in utilities savings in about a decade.

On seacoasts where the offshore and onshore wind currents are reliable along with other windy localities, utilities construct large windmill farms to generate electricity (Fig. 183). About 90 percent of the American wind power potential lies in 12 north-central and western states. However, even the most efficient wind farms cannot compete with the low prices for fossil-fuel-generated electricity. New designs, such as the highly efficient wind augmentation system using stacks of wind turbines, might make wind power a more economical alternative, with the added benefit of being nonpolluting.

Figure 181 *An artist's rendition of a solar electrical generation station.*

(Photo courtesy of U.S. Department of Energy)

Figure 182 *Solar panels on a laboratory at the University of California, Davis, California.*

(Photo courtesy U.S. Department of Energy)

Figure 183 *A wind-powered electrical generation station near Livermore, California.*

(Photo courtesy U.S. Department of Energy)

The wind also drives ocean waves, which could be harnessed to produce electricity. The breaking of a large wave on the coast is a vivid example of the sizable amount of energy that ocean waves produce. Many hydroelectric schemes have been developed to utilize this abundant form of energy, which is economical and efficient. Trapping tide waters in enclosed bays could be used to generate electricity by the power of falling water. An important use of falling water is hydroelectric dams. Nevertheless, hydroelectric dams are phenomenally expensive. The most accessible sites have already been exploited or else the flooding of large tracks of valuable land for new projects is unacceptable.

Pumped storage is another method of using water as a potential energy source. During off-peak hours, especially at night, when electricity consumption is down, an electric-motor-driven turbine pumps water up into a reservoir. In the daytime, during peak electrical demand, the water flows back through the pump, which operates as an electrical generator. This pumping-generating action smooths out the peaks and valleys during the generating cycle and allows power plants to operate at nearly full capacity, which greatly improves efficiency.

Ocean thermal-energy conversion, or OTEC, takes advantage of the temperature difference between thermal layers of the ocean to generate electricity. Surface water is boiled in a large, low-pressure steam generator. The water vapor is condensed using cold water brought up from great depths. The plants also produce freshwater as a by-product, another valuable resource. Many coastal areas around the world could take advantage of this unique form of solar energy.

The nutrient-rich coolant water could also be used for aquiculture, the commercial raising of fish, and serve nearby buildings with refrigeration and air-conditioning. The power plant could be located onshore, offshore, or on a mobile platform at sea. The electricity could supply a utility grid system, could be used on-site to synthesize substitute fuels such as methanol and hydrogen, could refine metals brought up from the seabed, or could manufacture ammonia for fertilizer.

Geothermal energy has an enormous potential. Much of the young mountain terrain in the western United States, as well as in Alaska and Hawaii, is of volcanic origin and forms a well-locked treasure of geothermal energy used for generating electricity. The potential geothermal energy resource in the United States alone is estimated at twice the energy of the world's petroleum reserves. Just a single eruption of Kilauea on the main island of Hawaii could supply two-fifths the power requirements of the entire United States during the time of the eruption.

In areas lacking natural geysers, geothermal energy can be extracted from fractured hot, dry rock in a method whereby water is injected into deep wells and steam is recovered. The dry, hot rock resources are several thousand times greater than all petroleum reserves. Hot, dry rocks lie beneath the surface in areas where the thermal gradients are two to three times greater than normal, about 100 degrees Celsius per mile of depth. The process of artificially making a geothermal reservoir within hot, buried rocks is difficult and expensive. If successful, though, the potential is enormous.

In a sense, Earth's interior can be thought of as a natural nuclear-power reactor because the heat is mainly derived by the decay of radioactive elements.

Figure 184 *A geothermal generating plant at the Geysers near San Francisco, California.* (Photo courtesy U.S. Department of Energy)

Many steam and geyser areas around the world are generally associated with active volcanism at plate margins. These are potential sites for tapping geothermal energy for steam heat and electrical power generation. Nations such as Iceland, Italy, Mexico, New Zealand, Russia, and the United States utilize underground supplies of superheated steam to drive turbine generators for electrical-power production.

Geothermal energy could prove to be far more valuable in the long run than petroleum, coal, or even nuclear energy. Besides, it is nonpolluting. Earth's internal heat will last for billions of years. Unlike limited resources of fossil fuels, geothermal energy has the potential of supplying people's energy needs for millennia. The geothermal resources of the United States alone are about 10 times the heat energy of all the nation's coal deposits. Unfortunately, overproduction of steam fields such as the Geysers in California (Fig. 184), the largest geothermal electrical generating plant in the world, could rapidly deplete this valuable natural resource.

After discussing the conservation of natural resources, the last chapter focuses on Earth's most valuable resources, namely the land and its life.

10

LAND USE

THE CHANGING LANDSCAPE

This chapter examines the land surface of Earth and how it is being used and misused. The most important environmental issue facing the world today is the appropriate use of the land and its natural resources to be preserved for future generations. Unfortunately, people have abandoned the concept of good stewardship of the land to provide for 1 billion more people added to the population rolls every decade. Certain environmental changes are occurring at rates never seen before in human history. Human activities are speeding up the rate of global change to such an extent they have attained the magnitude of a geologic force.

High population growth with its rising demands on the environment and increasing pollution is in the process of transforming the planet in a manner comparable to the effects of long-term geologic processes. Pollutants discharged into the air and water are permanently altering the biosphere and changing the global climate. Dramatic changes are occurring worldwide from the improper use of land and water resources, large-scale extraction and combustion of fossil fuels, widespread usage of chemicals in industry and agriculture, and global destruction of wildlife habitats. Thus, human beings constitute a major geologic force on the face of the planet.

GLOBAL ENVIRONMENTS

Earth's land surface comprises roughly one-third desert; one-third forests, savanna, and wetlands; and one-fifth glacial ice and tundra. The remaining land is inhabited by people. Deserts are the hottest and driest regions and among the most barren environments. In the Northern Hemisphere, a series of deserts stretches from the west coast of North Africa through the Arabian peninsula and Iran and on into India and China. In the Southern Hemisphere, a belt of deserts runs across South Africa, central Australia, and west-central South America.

Much of the world's desert wastelands receive only minor precipitation during certain seasons. Some regions have gone virtually without rain for years. Because of these forbidding conditions, desert areas cannot support significant human populations without artificial water supplies. About one-sixth of the human population lives in the drylands bordering the deserts in an area covering about one-quarter of Earth's landmass.

The world's tropical rain forests cover only about 7 percent of the land surface (Fig. 185). However, they contain two-thirds or more of all species. Plants and animals of the rain forests are being crowded out by human encroachment into their habitats. This is resulting in the destruction of ecological

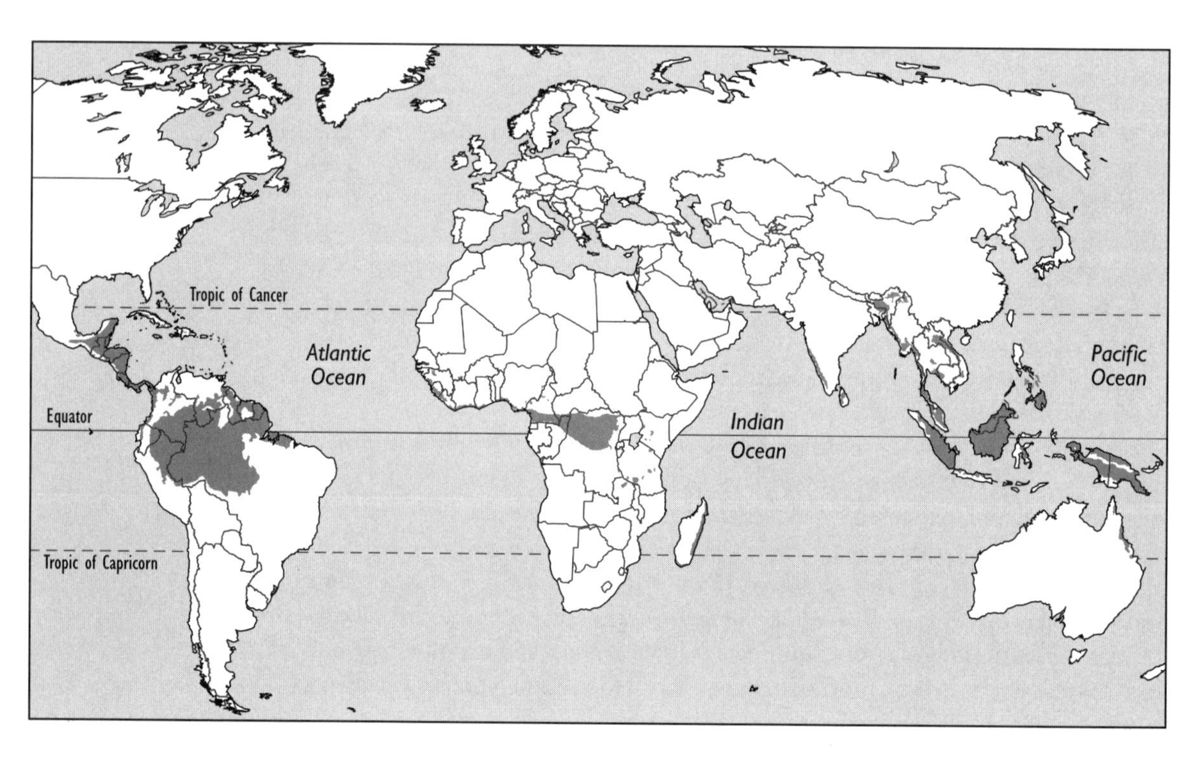

Figure 185 *The world's tropical rain forests.*

Figure 186 *An estuary of Twelvemile Creek, Niagara County, New York.*

(Photo by G. K. Gilbert, courtesy USGS)

niches and the pollution of the environment. Some exotic plants living exclusively in the tropical rain forests are rapidly dying out. When 90 percent of a forest habitat is lost, half its species of plants, animals, insects, and microbes are slated for extinction. High species diversity protected ecosystems against natural catastrophes, giving species-rich habitats a built-in mechanism to fight disasters. Conversely, ecosystems with low numbers of species are in danger of collapsing during bad times.

Half the rain forests of the world have already been destroyed for agriculture and timber harvesting. The remaining forests are in danger of deforestation as well. The great northern boreal forest is a vast band of conifers and other softwoods stretching across the northlands of North America and Eurasia. Over the past century, forests have declined due to logging and massive increases in tree dieback from fires, acid rain, and diseases, generally due to warmer weather in much of the region.

Wetlands (Fig. 186) are among the richest ecosystems in the world. They support many species of plants and animals, including valuable fisheries. About two-thirds of the shellfish harvested in the United States relies on these areas for spawning and nursery grounds. Wetlands also function as natural filters, removing sediments and some types of water pollution. Furthermore, they reduce flooding by absorbing excess runoff. They also protect coasts against storms and the serious erosion problems that accompany them.

The world's wetlands are rapidly disappearing, mostly due to destructive human activities. Wetlands are drained worldwide to provide additional farmland. The urgent need to feed growing populations is the major reason the developing countries drain wetlands. Short-term food production has obscured the long-term economic and ecological benefits of preserving wetland habitats. The disappearance of the wetlands is responsible for the loss of local fisheries and breeding grounds for marine species and wildlife. In many cases, wetland destruction is irreversible.

Many wetlands in the United States, such as the great Florida Everglades (Fig. 187), have been rapidly modified by human activities. Diking and filling of wetlands have eliminated habitats of fish and waterfowl. The introduction of exotic species has transformed the composition of aquatic communities. The reduction of freshwater inflow has changed the dynamics of plant and animal communities of the wetlands. In addition, urban and industrial wastes have contaminated sediments as well as organisms. The disposal of toxic wastes and the further reduction of freshwater inflows continue to alter wetland water quality and biological communities.

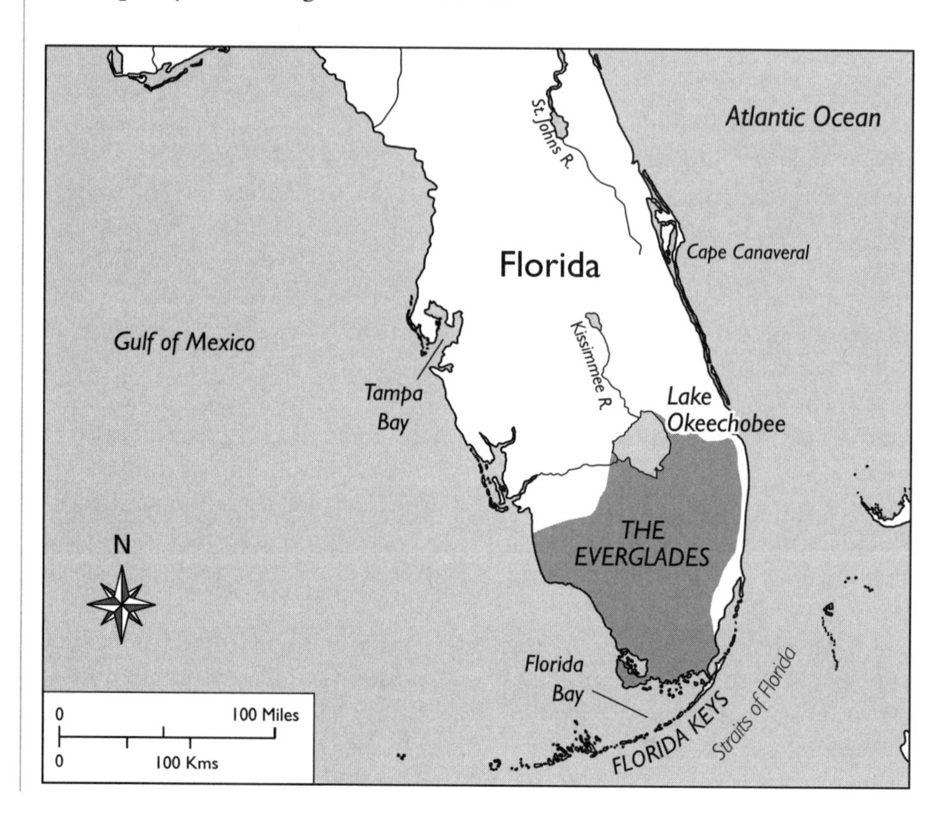

Figure 187 *The Everglades of southeast Florida are being rapidly modified by human activities.*

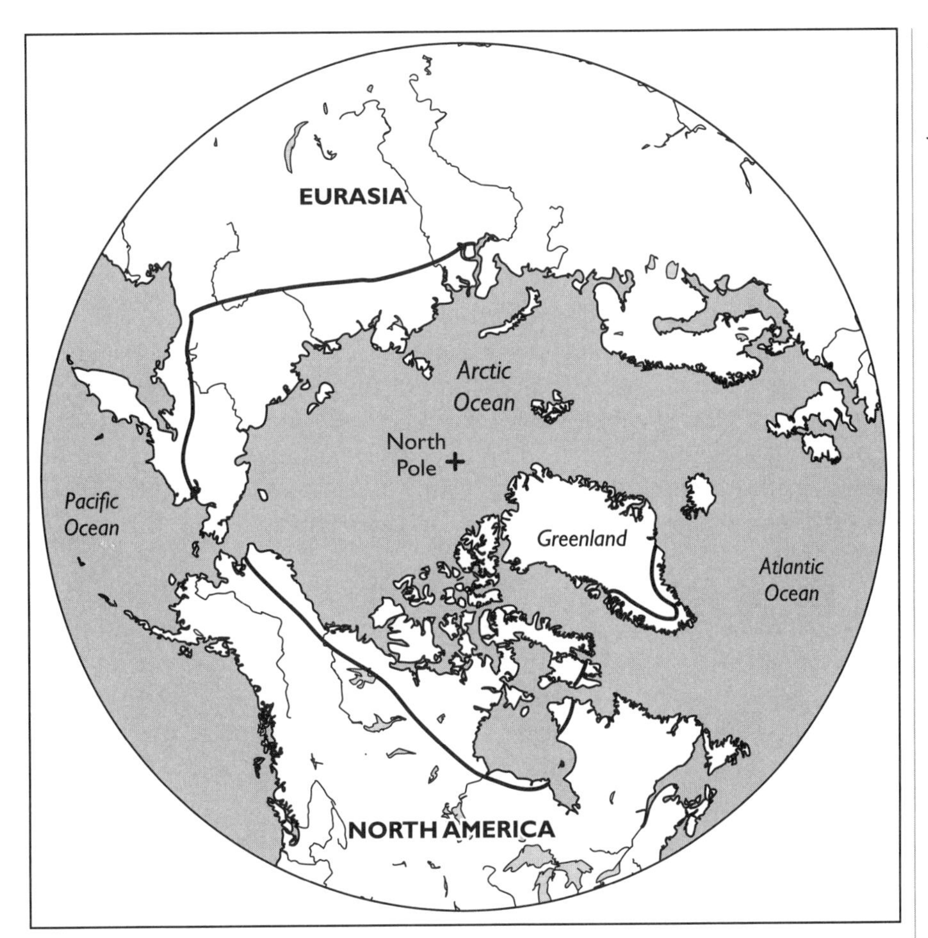

Figure 188 *The Arctic tundra line, north of which the ground remains frozen year-round.*

Nearly 90 percent of recent wetland losses in the United States have been for agricultural purposes. Woodland marshes are disappearing at an alarming rate of more than 1,000 acres a day. The numbers of migratory ducks and other waterfowl have fallen drastically due to drainage of their breeding grounds. In the last century, North America has lost 40 of the approximately 950 fish species. As sea levels continue to rise because of higher global temperatures, 80 percent of the coastal wetlands and estuaries could be lost by the middle of this century.

The Arctic tundra of Eurasia and North America (Fig. 188) covers about 14 percent of the world's land surface in an irregular band winding around the top of the world, north of the tree line and south of the permanent ice sheets. The Arctic surrounds the North Pole above 66.5 degrees north latitude. It embraces all the extreme northern lands around the Arctic Ocean, including the upper portions of Alaska, Canada, most of Greenland, the northern tip of Iceland, and the northlands of Scandinavia and Russia.

The climate in the Arctic varies more than anywhere in the world. Because Arctic tundra lies at such high latitudes, it is deprived of sunlight during the long winter months. The vegetation consists mostly of stunted plants often widely separated by bare rock and soil (Fig. 189). Limited food resources, high winds, and frigid temperatures during most of the year make the Arctic tundra one of the most barren regions on Earth.

Parts of the Arctic tundra are also the world's most nutrient-poor habitats. Special survival adaptations are required in this demanding environment, where sturdy cacti as well as frail insects live. In this land of unusual climatic conditions, plants and animals must take full advantage of the limited growing season, rainfall, and nutrients. The growing season is generally only two to three months long. However, a slight increase in temperature from global warming would extend the growing season, causing major changes in biologic communities.

The Arctic tundra is also among the most fragile environments. Even small disturbances can cause considerable damage. Overgrazing of reindeer on the sparse grasslands can decimate large areas. Petroleum and minerals exploration can ruin substantial acreage. Cross-country vehicle tracks remain decades later (Fig. 190). Arctic haze makes the region as polluted in winter and early spring as other places in the Northern Hemisphere afflicted with smog.

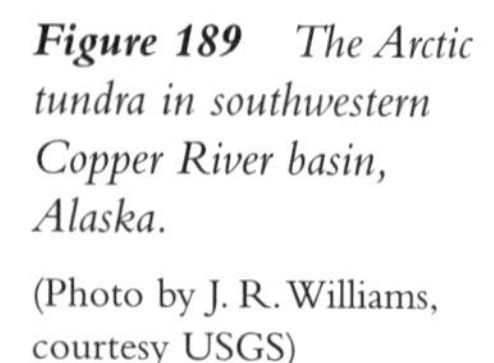

Figure 189 *The Arctic tundra in southwestern Copper River basin, Alaska.*

(Photo by J. R. Williams, courtesy USGS)

Figure 190 *Tractor trail on the North Slope of Alaska. The small ponds are due to thawing of the permafrost in the roadway.*

(Photo by O. J. Ferrians, courtesy USGS)

Coral reefs rank among the most biologically productive of all marine ecosystems. Corals are perhaps the busiest builders in the world. Their massive reefs upon which numerous communities depend for their survival even surpass the works of humans. Coral reefs are also centers of high biologic productivity. Their fisheries provide a major food source for the tropical regions. Unfortunately, the spread of tourist resorts along coral coasts in many parts of the world harms the productivity of these areas.

Developments along coral reefs are usually accompanied by increased sewage dumping, overfishing, and physical damage to the reef by construction, dredging, dumping, and landfills. Reefs are also destroyed to provide tourists with curios and souvenirs. In addition, upland deforestation chokes off coral reefs with eroded sediment carried by rivers to the sea. Nearshore species are particularly susceptible to coastal development, as human activity dramatically changes the nearshore habitats with increased pollution and sedimentation. Many coral reefs close to large human populations and home to a large number of marine species are in decline.

On many islands, such as Bermuda, the Virgin Islands, and Hawaii, urban development and sewage outflows have led to extensive overgrowth by thick mats of algae. These suffocate and eventually kill the coral by supporting the growth of oxygen-consuming bacteria. The reefs are particularly at risk during the winter, when the algal cover on shallow reefs is extensive (Fig. 191). This results in the loss of living corals and the eventual destruction of the reef by erosion.

The rise in ocean temperatures has caused the bleaching of many reefs. This has turned corals deathly white due to the expulsion of symbiotic (beneficial) algae from their tissues. The algae aid in nourishing the corals, providing as much as 60 percent of their food. Algal photosynthesis also speeds the growth of the coral skeleton by producing additional calcium carbonate. Therefore, the loss of algae poses a great danger to the reefs. Bleaching can also damage the coral's reproductive capacity, making recovery a long-term process if even possible. Foraminifera, marine plankton that are important players in the global carbon cycle and food chain, are suffering a similar bleaching effect. Other organisms that harbor algae in their tissues, including sea anemones, sea whips, and sponges, can also whiten in this manner.

Another disease attacking coral reefs near Key West, Florida, is called white pox. It manifests itself as discolored blotches and attacks the living tissue of the corals, which disintegrates and falls away, exposing the underlying skeleton. In some areas, the disease has killed between 50 and 80 percent of

Figure 191 *The coral-algae zone of a reef fringing Agana Bay, on the island of Guam.*

(Photo by J. T. Tracey Jr., courtesy USGS)

Figure 192 *A Coast Guard icebreaker clears a channel into McMurdo Sound, Antarctica.*

(Photo by M. Mullen, courtesy U.S. Navy)

the coral. If it spreads unchecked, it could ruin the famed reefs of the Keys and perhaps damage the entire marine ecosystem of the region. The corals act as coal miners' canaries. They reveal the health of the reef, which is home to fish and other animals that colonize the reef structure. Many developing countries depend heavily on coral reefs for their food supply. The destruction of the reef environment is a costly loss of nourishment.

Only 20 million square miles, about one-third of the planet's landmass, remains wild, with little signs of human perturbation. These signs include roads, settlements, buildings, airports, railroads, pipelines, power lines, dams, reservoirs, and oil wells. Other than a few scattered outposts around Antarctica (Fig. 192), the continent of ice is practically all wilderness. This situation could significantly change as nations begin exploring there for petroleum and mineral resources.

Several broad belts of wilderness wind around the globe. One band stretches across the Arctic tundra of northern Alaska, Canada, and the northernmost reaches of Eurasia. Another runs southwest from far eastern Asia through Tibet, Afghanistan, and Saudi Arabia into Africa. The Sahara Desert in northern Africa and Australia's great central desert are among the least densely populated regions on Earth.

Wild patches also exist in other parts of Africa, around the Amazon, and along the Andes Mountains of South America. Less than 20 percent of the identified wilderness areas are legally protected from exploitation. At least half the remaining wildlands are not self-protecting by virtue of their forbidding natures. They could be easily destroyed as billions of more people are added to the world's population.

DEFORESTATION

Deforestation results in large part from the consequences of poor forest management by governments with significant economic problems and rapidly growing populations. Forest clear-cutting is a tragic waste of a valuable natural resource. Impoverished countries cannot afford to lose such an important source of revenue. Many nations richly endowed with forests have created economic incentives to stimulate rapid depletion of their timber resources and the conversion of forestlands to agriculture and other uses. Less than 1 percent of the remaining tropical forests are being actively managed for sustained productivity.

Three-quarters of the global deforestation is conducted by landless, poverty-stricken people in a desperate struggle for survival. As much as 70 percent of the wood harvested in poor tropical countries is used locally for firewood. Because of the scarcity of firewood, the only source of fuel for heating and cooking in impoverished regions, their forests are rapidly being depleted. As the forests recede, severe firewood shortages loom ahead.

Only about 3.8 million square miles of tropical rain forest (an area about the size of the United States) remain in the entire world. Tropical rain forests in the Amazon Basin of South America are decreasing at an alarming rate at about 65 acres per minute or about 35 million acres annually, an area about the size of Arkansas. Throughout the world, billions of dollars worth of timber is simply burned. In the Amazon jungle of Brazil, some 20 million acres of forests are destroyed by fire each year. Developers have already slashed and burned some 20 percent of the Amazon rain forest (Fig. 193). If the rate of destruction continues, the forests will practically disappear by the middle of this century.

The tropical rain forest along the Atlantic coast of Brazil has dwindled to less than 1 percent of its original cover. In its place is a huge, man-made desert. The montane forests along the flanks of the Andes Mountains are also severely threatened, with those in the north already 90 percent deforested. Unlike the Amazon rain forests, the montane forests are extremely delicate. Pressures from a burgeoning human population threaten what little is left. Migration from rural areas to mountain cities has swelled over the past several decades. So today, more than 70 million people live in the northern Andes. People harvesting timber and clearing trees for roads, settlements, and agriculture are rapidly denuding the forests.

Parts of other continents are losing a greater percentage of their forests than is South America. More than 80 percent of Mexico's tropical rain forest has been destroyed. The forest cover of the Ivory Coast in western Africa has decreased by 75 percent since 1960. In terms of percentage of deforestation, continental Southeast Asia is losing a larger proportion of its forests each year. Indonesia is losing its rain forests, as much as 2,500 square miles a year, due to resettlement from the overcrowded main island of Java.

Figure 193 *The Amazon basin of South America is obscured by smoke from clearing and burning of the tropical rain forest, viewed from the space shuttle* Discovery *in December 1988.*

(Photo courtesy NASA)

The United States retains only about 15 percent of its once vast sea of forests. The remaining forests are seriously damaged by infestation or fires (Fig. 194) or are rapidly being depleted for timber products. About 50,000 acres of old-growth forests, mainly in the Pacific Northwest, are cut annually for lumber. Some areas in the Pacific Northwest that were clear-cut nearly a half century ago are still barren because of severe soil erosion. In the Appalachian Mountains, forests clear-cut nearly 100 years ago and allowed to grow back naturally have yet to return to their former condition and might never do so.

Figure 194 *Forest fires in Yellowstone National Park destroyed nearly half the forested land in the summer and fall of 1988.*

(Photo courtesy National Park Service)

A replanted forest requires about 200 years to restore the ecology of an old-growth forest, whose trees are among the oldest living things, often surviving for thousands of years. Less than 1 percent of the old-growth forests that once covered the eastern United States remain. The same prospect lies ahead for the old-growth forests of the Pacific Northwest and Alaska if trends continue.

The rain forests are being cleared mostly for agricultural purposes (Fig. 195). As developing nations attempt to raise their standards of living, one of the first steps is to clear forests and drain wetlands for agriculture. Much of the land is cleared by wasteful slash-and-burn methods. With these procedures, forests are set ablaze. The ashes fertilize the thin, nutrient-poor soil. Since artificial fertilizers are too expensive for farmers in the developing countries, the soil quickly wears out after a few years of agriculture. The fields are then abandoned, and more forests are put to the torch. The deserted farms are subjected to severe soil erosion due to the loss of vegetative cover that protects against the effects of wind and rain. The ground is laid bare to the elements. The soil is severely eroded, making forest recovery highly difficult.

Logging is often the first step toward deforestation, as loggers build roads into the forest, which paves the way for farmers. About 15 percent of the trees are cut down for timber production, much of which is wasted by inefficient harvesting and milling methods. The rapid decline of the rain forests is mostly a consequence of modern methods of timber harvesting, including the widespread use of chain saws and bulldozers. Lumber companies employ timber-harvesting equipment that snips trees off at the base with giant shears. Wood chippers can grind a 100-foot tree into pulp in seconds. After the most desirable trees are removed, unwanted trees and brush are ignited.

The soil underlying the rain forests is generally of poor quality. The fertilizing effect of the ash from burning trees is effective for no more than a few years because the nutrients are leached out by heavy rains. After intensive agriculture robs the soil of its nutrients, farmers are forced to abandon their fields and clear the forests for more land. When the rains return, flash floods wash away the denuded soil down to bedrock, and the rain forest has no chance of recovery. The destruction of large parts of the rain forests changes precipitation patterns, with the potential of turning wide areas into man-made deserts.

Soil erosion resulting from large-scale deforestation can overload rivers with sediments, causing considerable problems downstream. Monsoon

Figure 195 *Heavy haze over Zaire, Africa, created by agricultural burning, makes it impossible to see the ground from space shuttle* Challenger.

(Photo courtesy NASA)

floodwaters cascading down the denuded foothills of the Himalayas of northern India and carried to the Bay of Bengal by the Ganges and Brahmaputra Rivers have devastated Bangladesh, where several thousand people have lost their lives to floods. South America's Amazon River is forced to carry more water during the flood season due to deforestation at its headwaters. Deforestation has a severe environmental impact on soil, water quality, and local climate. Fisheries in rivers and lakes are damaged by increased sedimentation from soil erosion in deforested areas.

Deforestation is also contributing to the rise in sea levels. The extraction of groundwater, redirection of rivers for agriculture, drainage of wetlands, deforestation, and other activities that divert water to the oceans account for about one-third of the global sea level rise. When water stored in aquifers, lakes, and forests is released at a faster rate than it is replaced, the water eventually ends up in the oceans. Forests store water in both their living tissues and the moist soil shaded by plant cover. Also, one of the products of combustion when forests are burned is water. When forested areas are destroyed, the water within eventually winds up in the ocean, thus raising the sea level.

HABITAT DESTRUCTION

Earth is in danger of losing the battle for its forests and wildlife habitats. More than 90 percent of all species occupy the land due to its large number of ecosystems. For the first time in geologic history, plants are being extinguished in tragic numbers. If current trends continue, a significant number of plant species are likely to become extinct. In the United States alone, 7 percent of the nation's plant species are destined for extinction. Plants are at risk of extinction from forest destruction, expansion of agriculture, and the spread of urbanization.

More than 1,000 domestic species of plants and animals are either endangered or threatened with extinction. Possibly by the middle of this century, the number of extinct species could exceed those lost in the great extinctions of the geologic past. Plants and animals are forced into extinction as growing human populations continue to squander Earth's space and resources and to contaminate the soil, water, and air.

Valuable plant and animal species are dying out at alarming rates due to human encroachment onto wildlife habitats. The species threatened with extinction include about 35 percent of fish, 25 percent of amphibians, 25 percent of mammals, 20 percent of reptiles, and 10 percent of birds. Species are disappearing mainly due to deforestation in the tropics and elsewhere in the world. Freshwater fish species are rapidly disappearing worldwide from deforestation, which increases sedimentation, and from acid rain, which acidifies streams and lakes (Fig. 196).

Figure 196 *A researcher from the Forest Service tests a lake for acidity levels.*

(Photo courtesy U.S. Forest Service)

Rain forests are the homes of about 80,000 plant species. Certain exotic plants are rapidly disappearing. A large percentage of plant species are likely to become extinct if current trends continue. Some plants have an important medicinal value. Half of all pharmaceuticals are manufactured from natural herbs, most of which live only in the tropical rain forests. Therefore, as the destruction of these forests continues, humanity loses the ability to find new cures for fatal diseases. Plants and animals of the rain forests are being crowded out by human encroachment into their habitats, resulting in the destruction of ecological niches and pollution of the environment.

By being aware of the bleak future that awaits many species, several nations have set aside game preserves in an attempt to halt the tide of habitat destruction and extinction. Yet even these areas represent less than 1 percent of the remaining forests. Africa, once a sea of wild animals surrounding a few islands of humanity, now has only a few enclaves of animals surrounded by a mass of people. The amount of forested land in the United States and a few other countries has actually increased slightly in recent years. The United States Forest Service has taken millions of acres of forestlands out of multiple use and established wilderness areas. Unfortunately, the forests surrounding these enclaves are still in danger of being destroyed.

African elephants disappear rapidly when the human population grows to a certain threshold. Since elephants are large animals with large needs, the main threats to their survival are land use clashes with people, causing them to migrate to other areas. These large herbivores actually improve their environment by opening forests for grass undergrowth, which increases productivity and accelerates nutrient recycling. The cleaner forests are also much less vulnerable to forest fires. Unfortunately, with the elimination of these animals, their favorable environmental impacts are reversed, restricting the habitats of smaller herbivores, which follow their larger cohabitants into extinction.

If global warming becomes too abrupt, forests, especially game preserves, could become isolated from their normal climate regimes, which continue to move to higher latitudes. Forests would creep poleward. Other wildlife habitats, including the Arctic tundra, would disappear entirely. Plants would be hardest hit because they are directly affected by changes in temperature and rainfall. Many species would be unable to keep pace with these rapid climate changes. Those that are able to migrate could find their routes blocked by natural and man-made barriers, including cities and farms. The climate change could rearrange entire biological communities and cause many species to become extinct. Others, commonly called pests, would overrun the landscape.

The disappearance of the rain forests could cause a decline in bird populations for the northern countries because they are the wintering grounds for migratory species. Countless other species that inhabit the forests themselves are also dying out. Already, songbirds are disappearing in tragic numbers. Birds are particularly at risk. Humans have forced a large number of bird species into extinction over the past centuries. About one-fifth of the present bird species are endangered or near extinction from human activities that have directly or indirectly altered the environment in a manner detrimental to birds.

As a possible prelude to global extinction is the alarming disappearance throughout the world of frogs and other amphibians that have been living on Earth for more than 300 million years. Amphibians such as frogs have developed deformities, including multiple or missing legs, possibly caused by pollutants such as pesticide and fertilizer runoff. As with all amphibians, frogs have permeable skins that can absorb toxins from the environment. Since the 1960s, due to deforestation, acid rain, pollution, or ozone depletion, frog species have been going extinct in large numbers. Furthermore, amphibians are vanishing from nature preserves, where little human perturbation occurs. For instance, in Costa Rica's Cloud Forest Preserve, 20 out of 50 once abundant species of frogs have not been seen in several years. The deaths of these creatures might be sounding an early warning that the planet is in grave danger.

LAND ABUSE

Land use has long been suspected of changing the climate as well as the environment. The climatic effects of deforestation, grazing, agriculture, and development have profoundly altered vast areas of land on almost every continent. People are having a huge impact on the environment through local and regional land practices. Human activities over large parts of the world are seriously disrupting patterns of land and water use. Global destruction of forests and wildlife habitats, large-scale extraction and combustion of fossil fuels, and use of toxic chemicals in industry and agriculture are permanently altering cycles of essential nutrients in the biosphere.

These activities might also affect the global climate and change precipitation patterns, causing droughts and loss of farm productivity (Fig. 197) at a time when it is most needed. The effects of droughts are steadily worsening because of the deepening poverty, increasing population, and the abuse of the land. Land use changes are also altering the hydrologic cycle, causing a permanent decrease in rainfall and soil moisture.

People in many parts of the world are unable to feed themselves properly due to the destruction of their land. The loss of agricultural production is the result of vanishing forests and wetlands, topsoil depletion and desertification, improper irrigation methods and overuse of groundwater, population pressures on limited food and natural resources, and the effects all these interrelated problems have on political and economic stability.

Maintaining food production while simultaneously destroying the very land needed to feed future generations is senseless. Many nations just manage

Figure 197 *A dust storm on a farmstead in Baca County, Colorado, during the 1930s Dust Bowl years. The soil has become airborne, forming clouds of dust so dense that visibility at times is zero and soil drifts around structures.*

(Photo courtesy USDA Soil Conservation Service)

to feed themselves. However, they achieve this effort at the expense of their soil and groundwater, thereby making long-term survival doubtful. Each year, the world's farmers must feed an additional 100 million people on 20 billion fewer tons of topsoil. People in many parts of the world are unable to feed themselves properly because they have destroyed their land. Nations are caught in a quandary over whether they should feed their growing populations today or save the land for tomorrow.

About 10 percent of the land is used for farming and about 25 percent for pasture. Some 400 square miles of arable land are required to feed 1 million people. A doubling of the human population predicted by the middle of this century would require that either more land be placed under cultivation or twice as much food be grown on the existing land. In the former case, more forests would have to be cleared and wetlands drained. In the latter case, such intensive agriculture would ultimately destroy the land.

Under increasing pressure for more food production, normally fallow fields are cultivated, which quickly wears out the soil. The United States no longer has an excess capacity in such basic agricultural resources as land, water, and energy. A shortage of petroleum would result in lower amounts of nitrogen-based fertilizers as well as fuel for running farm machinery and pumping irrigation water. Efforts to farm the weak soils of the rain forests have been disastrous. Overirrigation is destroying substantially large acreages due to salt accumulating in the soil, which is becoming one of the greatest factors limiting agricultural productivity.

Figure 198 *America has most of its arable land in production as shown here by these loess uplands abutting farmland near Wauneta, Nebraska.* (Photo by W. D. E. Cardwell, courtesy USGS)

The world's leading food exporters already have most of their arable land in production (Fig. 198). To increase export, American farmers have put into production millions of additional acres. Many of these are substandard, including sloping, marginal, and fragile soils that are vulnerable to erosion. As a result, the United States has actually lost agricultural land. Every year, expanding urbanization removes another 1 million acres of valuable cropland.

URBANIZATION

Nearly half the world's people live in cities. Four-fifths of the population growth in the 1990s has been in urban areas, creating unprecedented concentrations of people. High population density usually indicates good climatic and growing conditions, with no deserts, mountains, or other impediments to human habitation. The countries that are least populated often have poor climates or uninhabitable geographies. About one-third of the world's landmass is composed of uninhabited wilderness areas. However, only half these regions are self-protected by virtue of their forbidding nature.

The most densely populated regions are islands and low-lying river deltas. The tiny island of Macau located off the coast of China west of Hong Kong is the most densely populated place on Earth. Some 350,000 people are jammed onto only 6 square miles, resulting in about 60,000 people per square mile. If the population were spread evenly throughout the island, one person would occupy the space of one-quarter of a tennis court. Bangladesh has about 160 million people squeezed into an area about the size of Wisconsin, with nearly 2,000 people per square mile.

Overcrowding is wrecking the very fabric of society. It is causing rampant unemployment, lawlessness, homelessness, and a host of other social ills. Overcrowding and growing scarcities of valuable natural resources are contributing to violent conflicts in many parts of the world. This condition is especially true for poor countries, where shortages of forests, agricultural land, and water resources along with rapidly expanding populations are causing unbearable hardships.

Every society attempts to provide its people with the basic requirements of life, including adequate food and shelter along with a healthy environment. Only when these essential needs are assured can attention be turned to comfort and convenience, which determines a society's quality of life or living standard. Unfortunately, for much of the world, the quality of life suffers as populations continue growing well beyond the capacity of their land to provide for them. As a result, people are forced to spend more time and effort obtaining enough food to stay alive, with little income for improving their standards of living.

However, even a stationary population that continues to improve its standard of living by increasing demands on natural resources is just as destructive to the environment as a growing population with slow increases in living standards. Environmental protection is becoming exceedingly difficult in the face of increasing global populations, increasing concentrations of populations in cities, and rising standards of living, which rely on large inputs of natural resources.

Overcrowded cities are more vulnerable to natural disasters. An estimated 25,000 people have died in natural disasters since 1975, and four times as many were injured. These events caused damages amounting to 500 billion dollars, which does not take into account indirect costs such as lost business and employment, environmental harm, and emotional tolls on victims. Severe storms caused most of the losses, while earthquakes and volcanoes accounted for the rest. Seven of the 10 most expensive disasters have occurred since 1989, indicating the cost of catastrophes is growing.

People crowding onto coastal regions and low-lying river deltas are particularly at risk from tropical storms. Earthquakes take the lives of a great many people, as buildings topple down upon them (Fig. 199). Population growth in the past decades has pushed people into closer proximity of the world's 600 active volcanoes. During the last century, volcanoes have killed on average

Figure 199 *Rubble from a brick building that completely collapsed during the July 28, 1976, Tangshan, China, earthquake that killed 250,000 people.*

(Photo courtesy USGS)

some 1,000 people annually. This number is bound to rise as populations living in the domain of dangerous volcanoes continue to increase.

Twenty cities, mostly in the undeveloped countries, contain populations exceeding 10 million people each. Many cities are becoming sprawling slums, with few services and much disease, pollution, crime, unemployment, and political unrest. Epidemics are on the rise around the world, especially in undeveloped countries where expanding urban slums pose a serious health hazard.

Since the dawn of civilization, people have settled in valleys, and wars have been fought over river courses. Factories were located near rivers because waterways provided easy transport of materials, sufficient water for processing and cooling, and convenient disposal of wastes into the river itself. When populations were sparse and industrialization was in its infancy, the dilution of industrial pollution by river water had little environmental impact. However, today, with exploding human populations and supportive industries, pollution has become a serious environmental problem. Although valleys are generally the preferred sites for industrialization, they are also more likely to experience temperature inversions that trap air pollution.

More than three-quarters of the American people live in or near metropolitan areas with populations exceeding 50,000. This shift in demographics, due to a transition from an agrarian to an industrial economy, is a complete reversal compared with a century ago, when half the people still lived and worked on farms. At the same time in Europe, with the Industrial Revolution gaining momentum, populations were becoming more urban than rural. The lure of big cities and the prospect of good jobs prompted a mass migration from farm to factory.

The evolution of urban areas has been from farms to small towns to cities to large metropolitan areas to urban regions with at least 1 million people. In 1920, the major urban regions in the United States contained about one-third of the total population of the country, which was mostly rural. Fifty years later, that situation completely reversed, with urban regions containing most of the population. Presently, the major urban regions contain more than 80 percent of all Americans. Moreover, sprawling urban regions occupy one-sixth the total land area of the continental United States. Accompanying this urban sprawl are serious environmental impacts. For example, a rise in automobile traffic associated with urban sprawl has led to an increased rain of pollutants entering local waters.

Expanding urbanization removes millions of acres of valuable cropland every year. The United States is losing prime farmland at a rate of 50 acres per hour. Urban sprawl is engulfing adjacent farms so fast that the country might be forced to import food if agricultural land continues to disappear at such a rapid rate while the population continues a rapid upward expansion. The

Figure 200
Unprotected cropland in South Dakota erodes, washing away topsoil and polluting streams with sediments.

(Photo by Tom Pozarnsky, courtesy USDA Soil Conservation Service)

population is expected to grow by 50 percent. Farmland, however, is projected to shrink by as much as 15 percent by the middle of the century.

The destruction of the best agricultural land places greater pressure onto less productive and more environmentally fragile lands. This leads to excessive soil erosion (Fig. 200). The conversion of rural land to nonagricultural uses in the United States amounts to about 2 million acres annually. The rural land is converted to urban development, transportation facilities, reservoirs, wildlife refuges, wilderness areas, parks, and recreational areas. About 3 percent of the land area is covered by buildings and roads.

Urbanization makes much of the soil within the urbanized area impervious to rainfall and runoff. Therefore, drainage water overflows into the streets, causing localized flooding when sewers cannot handle the excess (Fig. 201). The urban environment has increased the magnitude and frequency of floods in small drainage basins. The rate of increase is determined by the amount of land covered by impenetrable pavement and cement and by the area served by storm drainages.

Growing cities sacrifice a significant number of trees to development. Their loss contributes to the so-called heat-island effect, which makes pavement and buildings heat reservoirs that warm urban areas several degrees more than the surrounding countryside. The heat-island effect created by city buildings and pavement along with the buildup of greenhouse gases from industrialization appears to be leading to global warming. The heat gain in the cities requires additional artificial air-conditioning to replace the natural cooling effects of vegetation, placing higher demands onto energy resources. The loss of trees also reduces the absorption of excess carbon dioxide and other pollutants generated within the city itself.

While alternative nonpolluting energy sources are being developed, the carbon dioxide buildup in the atmosphere could be curtailed by planting trees. By doubling the volume of forest growth each year, the major fossil-fuel-consuming nations could delay the onset of global warming by perhaps a decade or more. However, the destruction of the tropical rain forests would have to be halted as well. Cutting old-growth forests adds carbon dioxide to the atmosphere by the combustion, decomposition, and processing of wood products, which contributes to the greenhouse effect.

A forest covering nearly 3 million square miles, an area roughly the size of the United States, would be required to restore fully Earth's carbon dioxide

Figure 201 *Damage to homes in Grantsville, West Virginia, caused by flooding of the Little Kanawha River on March 6, 1967.*

(Photo by E. A. Gaskins, courtesy USDA Soil Conservation Service)

balance. This equals an area of all tropical forests cleared since the dawn of agriculture. By planting additional trees, enough forest growth could be added to absorb excess carbon dioxide being discharged into the atmosphere by human activities. Replanting perhaps as many as 100 million trees would remove about 18 million tons of carbon dioxide from the atmosphere each year. New trees require two to three decades to mature. Furthermore, immature trees do not absorb as much carbon dioxide as those they replace. For much of the world, deforestation has destroyed the topsoil, so replanting trees is no longer an option. However, new forests could be planted on many degraded lands that are no longer agriculturally productive (Fig. 202).

LAND USE PLANNING

Modern technology, including vast arrays of satellites and powerful computers, can provide the measurement and computation tools to study Earth as a complete system. Scientists have gained comprehensive knowledge of the state of the planet and of its global processes. They also have become uncomfortably aware that major changes are taking place and that humans are responsible for serious disruptions to the planet. If the world's population grows as

Figure 202 *The ruins of an old cabin and deep, eroded gulleys on bared slopes below spruce woods, Carter County, Tennessee.*

(Photo by A. Keith, courtesy USGS)

Figure 203 *An abandoned quarry that has been converted into a park for water recreation.* (Photo by E. A. Imhoff, courtesy USGS)

predicted and human activity remains unchanged, science and technology might be powerless to prevent widespread poverty and irreversible damage to the environment. When the world is filled with too many people, technology can no longer solve the problems but can only postpone them to a future date, at which time they could spread well beyond people's control.

Tropical rain forests in the Western Hemisphere once covered 3 million square miles. That area has now dwindled to about one-third of its original cover. In addition, as much as three-quarters of the tropical rain forests of Africa have been destroyed. The detection, identification, and measurement of the symptoms of forest decline using satellite technology is urgently needed. Such observations can provide a means to assess and monitor forest destruction on a global scale, giving governments critical information to help reduce forest and wildlife habitat destruction.

Not all land is the same. Its particular physical and chemical characteristics might be more important to urban development than its geographic location alone. The supply of land is also limited. Therefore, land use planning is needed to make suitable land available for specific uses. The need for additional land near urban areas has led to the reclamation of land formally used for other purposes such as landfills or mining (Fig. 203).

The landscape is evaluated for land use planning, site selection, construction, and the environmental impact related to these activities. The role of geology in landscape evaluation is to provide surface and subsurface geologic information needed for evaluation, design, and construction of projects such as buildings, highways, airports, reservoirs, tunnels, pipelines, and recreational facilities.

Specific information used for landscape evaluation generally includes the physical and chemical properties of earth materials, bedrock depth, slope stability, seismic risk assessment, groundwater characteristics, and the presence of floodplains. Without proper information on the geology of the land, haphazard development during rapid growth could lead to death and destruction from natural disasters that have always plagued the area.

As forested and agricultural lands give way to continued urbanization, the change in spectral characteristics is readily discernable on satellite imagery. Up-to-date imagery of land use from satellites is becoming increasingly necessary for planners to attempt to solve the ever-growing problems of overpopulation on a planet with finite resources. The gatherings of accurate data, including that provided by satellites, along with intelligent planning will ensure the best-possible use of valuable resources.

In urban areas, satellite imagery can identify different levels of habitation such as the central business district with a high density of buildings, dense residential areas with grass cover, and sparse residential areas with a moderate growth of trees. In many large American cities, a central core is surrounded by a prosperous and growing suburban and exurban region in a so-called doughnut complex.

Built-up areas generally appear blue gray on multispectral satellite imagery because of the spectral signature of concrete, which tends to absorb near-infrared radiation on the low end of the solar spectrum. Grass yards and stands of trees appear red because of their high reflectance of near-infrared radiation. The imagery, called false-color photography, is extremely useful for mapping vegetation. Comparing imagery taken at various times can therefore monitor urban sprawl as metropolitan areas continue to expand their city limits into the once pristine countryside.

CONCLUSION

A final word must be said about the effects the human population explosion has on the environment. The human population explosion is too often missing from the debate on the degenerating condition of the planet. Poverty, environmental degradation, and other serious problems dealing with rapid population growth can no longer be ignored. Populations are growing so explosively and modifying the environment so extensively that people are inflicting a global impact of unprecedented dimensions. Large population increases could be disastrous, considering the destructive impacts of today's level of human activities. Global warming, pollution, acid precipitation, ozone depletion, deforestation, soil erosion, desertification, species extinction, and a host of other serious problems are a direct result of increased human activities.

The continuation of these activities could alter the balance of nature. The complicated interdependence organisms have on each other and on their environment is not yet fully understood. What is becoming more apparent, however, is that if humans continue to upset nature through wanton negligence and waste, future generations would be left with an entirely different biological world than the one inhabited today. As human populations continue to grow out of control, other species are forced aside to make room for additional agriculture, industry, and urbanization, along with the habitat destruction that accompanies these activities.

The human race is on a collision course caught between limited resources and the growing numbers of people using them. Rapid population growth is stretching the resources of the world. The prospect of future increases raises doubts whether the planet can continue to support people's growing needs without serious, irreversible damage. The depletion of natural resources could also jeopardize further human advancement. The increase in world economic activity needed to keep pace with rising human requirements could subject the biosphere to conditions it cannot possibly tolerate without irreversible damage. Humans are destroying the world's forests and pumping pollutants into the air and water, thus unfavorably changing the composition of the biosphere.

Human activities appear to be responsible for the many climatic disturbances that beset the planet. The composition of the atmosphere has changed significantly faster than at any other time in human history. In effect, the human race is conducting a dangerous, inadvertent global experiment by altering the environment with waste products. Every ton of carbon dioxide, every gallon of pollution, and every extinction brings the world closer to a habitability crisis. With more people in the world, more forests are cleared, more firewood is gathered, more topsoil is eroded, and more pollution is produced. If human population growth is not brought under control, nature will control it for us.

GLOSSARY

aa lava (AH-ah) a lava that forms large jagged, irregular blocks
abrasion erosion by friction, generally caused by rock particles carried by running water, ice, and wind
abyss (ah-BIS) the deep ocean, generally over a mile in depth
acid precipitation any type of precipitation with abnormally high levels of sulfuric and nitric acids
aerosol a mass of fine solid or liquid particles dispersed in air
agglomerate (ah-GLOM-eh-ret) a pyroclastic rock composed of consolidated volcanic fragments
air pollution the contamination of the air by natural and industrial activities
albedo the amount of sunlight reflected from an object and dependent on the object's color and texture
alluvium (ah-LUE-vee-um) stream-deposited sediment
alpine glacier a mountain glacier or a glacier in a mountain valley
aquifer (AH-kwe-fer) a subterranean bed of sediments through which groundwater flows
ash fall the fallout of small, solid particles from a volcanic eruption cloud
asperite (AS-per-ite) the point where a fault hangs up and eventually slips, causing earthquakes
asteroid a rocky or metallic body whose impact on Earth creates a large meteorite crater
asthenosphere (as-THE-nah-sfir) a layer of the upper mantle from about 60 to 200 miles below the surface that is more plastic than the rock above and below and might be in convective motion

atmosphere a thin membrane of gases divided into the troposphere, which ranges from 0 to 10 miles altitude and comprises 80 percent of the air mass, and the stratosphere, which ranges from 10 to 40 miles altitude and contains low-pressure stable air

atmospheric pressure the weight per unit area of the total mass pressure of air above a given point, also called barometric pressure

avalanche (AH-vah-launch) a slide on a snowbank triggered by vibrations from earthquakes and storms

back-arc basin a seafloor-spreading system of volcanoes caused by extension behind an island arc that is above a subduction zone

barrier island a low, elongated coastal island that parallels the shoreline and protects the beach from storms

basalt (bah-SALT) a dark volcanic rock that is usually quite fluid in the molten state

bedrock solid layers of rock beneath younger materials

bicarbonate an ion created by the action of carbonic acid on surface rocks; marine organisms use the bicarbonate along with calcium to build supporting structures composed of calcium carbonate

biodegradable capable of being broken down into environmentally safe substances by the action of living organisms

biogenic sediments composed of the remains of plant and animal life such as shells

biomass the total mass of living organisms within a specific habitat

biosphere the living portion of Earth that interacts with all other biologic and geologic processes

black smoker superheated hydrothermal water rising to the surface at a midocean ridge; the water is supersaturated with metals, and when exiting through the seafloor, the water quickly cools and the dissolved metals precipitate, resulting in black, smokelike effluent

blowout a hollow caused by wind erosion

blue hole a water-filled sinkhole

bomb, volcanic a solidified blob of molten rock ejected from a volcano

calcite a mineral composed of calcium carbonate

caldera (kal-DER-eh) a large, pitlike depression at the summits of some volcanoes and formed by great explosive activity and collapse

calving formation of icebergs by glaciers breaking off upon entering the ocean

carbonaceous (KAR-beh-NAY-shes) a substance containing carbon, namely sedimentary rocks such as limestone and certain types of meteorites

carbonate a mineral containing calcium carbonate such as limestone

carbon cycle the flow of carbon into the atmosphere and ocean, the conversion to carbonate rock, and the return to the atmosphere by volcanoes

carcinogen any natural or human-made substance that, when found in the environment at certain levels, causes cancer

catchment area the recharge area of a groundwater aquifer

circum-Pacific belt active seismic regions around the rim of the belt Pacific plate coinciding with the Ring of Fire

climate the average course of the weather for a certain region over time

coal a fossil-fuel deposit originating from metamorphosed plant material

coastal storm a cyclonic, low-pressure system moving along a coastal plain or immediately offshore

condensation the process whereby a substance changes from the vapor phase to the liquid or solid phase; the opposite of evaporation

conduit a passageway leading from a reservoir of magma to the surface of Earth through which volcanic products pass

cone, volcanic the general term applied to any volcanic mountain with a conical shape

contaminant any substance that pollutes the environment

continent a landmass composed of light, granitic rock that rides on the denser rocks of the upper mantle

continental drift the concept that the continents have been drifting across the surface of Earth throughout geologic time

continental glacier an ice sheet covering a portion of a continent

continental margin the area between the shoreline and the abyss that represents the true edge of a continent

continental shelf the offshore area of a continent in a shallow sea

continental shield ancient crustal rocks upon which the continents grew

continental slope the transition zone from the continental shelf to the deep-sea basin

convection a circular, vertical flow of a fluid medium by heating from below; as materials are heated, they become less dense and rise, cool, become more dense, and sink

convergent plate boundary the boundary between crustal plates where the plates come together; generally corresponds to the deep-sea trenches where old crust is destroyed in subduction zones

coral any of a large group of shallow-water, bottom-dwelling marine invertebrates that are reef-building common colonies in the Tropics

core the central part of Earth, consisting of a heavy iron-nickel alloy

Coriolis effect the apparent force that deflects the wind or a moving object, causing it to curve in relation to the rotating Earth

crater, volcanic the inverted conical depression found at the summit of most volcanoes, formed by the explosive emission of volcanic ejecta

creep the slow flowage of earth materials

crevasse (kri-VAS) a deep fissure in the crust or a glacier

crust the outer layers of a planet's or a moon's rocks

crustal plate a segment of the lithosphere involved in the interaction of other plates in tectonic activity

deforestation the clearing of forests for agriculture and other purposes

delta a wedge-shaped layer of sediments deposited at the mouth of a river

density the amount of any quantity per unit volume

desertification (di-zer-te-fa-KA-shen) the process of becoming arid land by natural processes or by mismanagement

desiccated basin (de-si-KAY-ted) a basin formed when an ancient sea evaporated

developed nation a heavily industrialized, generally rich country

dew liquid-water droplets formed by condensation of water vapor from the air as a result of radiation cooling

dew point the temperature to which air, at a constant pressure and moisture content, must be cooled for saturation to occur

diapir (DIE-ah-per) the buoyant rise of a molten rock through heavier rock

dike a tabular intrusive body that cuts across older strata

divergent plate boundary the boundary between crustal plates where the plates move apart; it generally corresponds to the midocean ridges where new crust is formed by the solidification of liquid rock rising from below

downwelling the sinking of a fluid that is heavier than the surrounding medium

drought a period of abnormally dry weather sufficiently prolonged for the lack of water to cause serious deleterious effects on agricultural and other biological activities

drumlin a hill of glacial debris facing in the direction of glacial movement

dune a ridge of windblown sediments usually in motion

earth flow the downslope movement of soil and rock

earthquake the sudden rupture of rocks along active faults in response to geologic forces within Earth

ecology the interrelationship between organisms and their environment

ecosystem a community of organisms and their environment functioning as a complete, self-contained biological unit

effluent an outflow of liquid waste material and usually considered a pollutant

elastic rebound theory the theory that earthquakes depend on rock theory elasticity

environment the complex physical and biological factors that act on an organism to determine its survival and evolution

eolian (EE-oh-lee-an) a deposit of windblown sediments

epicenter the point on Earth's surface directly above the focus of an earthquake

erosion the wearing away of surface materials by natural agents such as wind and water

estuary a tidal inlet along a coast and an important environment for fish and shellfish

evaporation the transformation of a liquid into a gas

evaporite (ee-VA-per-ite) the deposition of salt, anhydrite, and gypsum from evaporation in an enclosed basin of stranded seawater

evolution the tendency of physical and biological factors to change with time

exfoliation (eks-FOE-lee-A-shen) the weathering of rock causing the outer layers to flake off

extinction the loss of large numbers of species over a short geologic time

extrusive (ik-STRU-siv) an igneous volcanic rock ejected onto the surface of a planet or moon

fault a break in crustal rocks caused by earth movements

fauna the animal life of a particular area or age

fissure a large crack in the crust through which magma might escape from a volcano

floodplain the land adjacent to a river that floods during river overflows

flora the plant life of a particular area or age

fluvial (FLUE-vee-al) pertaining to being deposited by a river

fossil any remains, impression, or trace in rock of a plant or animal of a previous geologic age

fossil fuel an energy source derived from ancient plant and animal life and includes coal, oil, and natural gas; when ignited, these fuels release carbon dioxide that was stored in Earth's crust for millions of years

frost heaving the lifting of rocks to the surface by the expansion of freezing water

fumarole (FUME-ah-role) a vent through which steam or other hot gases escape from underground such as a geyser

geologic column the total thickness of geologic units in a region

geomorphology (JEE-eh-more-FAH-leh-jee) the study of surface features of Earth

geothermal the generation of hot water or steam by hot rocks in Earth's interior

geyser (GUY-sir) a spring that ejects intermittent jets of steam and hot water

glacier a thick mass of moving ice occurring where winter snowfall exceeds summer melting

glacier burst a flood caused by an underglacier volvanic eruption

glacière (GLAY-sher-ee) an underground ice formation

graben (GRA-bin) a valley formed by a downdropped fault block

gravity fault motion along a fault plane that moves as if pulled downslope by gravity; also called a normal fault

greenhouse effect the trapping of heat in the lower atmosphere principally by water vapor and carbon dioxide

groundwater water derived from the atmosphere that percolates and circulates below the surface

guyot (GEE-oh) an undersea volcano that reached the surface of the ocean, whereupon its top was flattened by erosion; later, subsidence caused the volcano to sink below the surface

haboob (hey-BUBE) a violent dust storm or sandstorm

hazardous waste any pollutant that is particularly harmful to life, including toxic substances and nuclear wastes

heat budget the flow of solar energy through the biosphere

heat flow heat energy transfers from hot toward cold at a rate or flux equal to the temperature gradient times the conductivity of the material in between

hot spot a volcanic center with no relation to a plate boundary; an anomalous magma generation site in the mantle

hydrocarbon a molecule consisting of carbon chains with attached hydrogen atoms

hydrologic cycle the flow of water from the ocean to the land and back to the sea

hydrology the study of water flow over Earth

hydrosphere the water layer at the surface of Earth

hydrothermal relating to the movement of hot water through the crust; also a mineral ore deposit emplaced by hot groundwater

hypocenter the point of origin of earthquakes; also called focus

ice age a period when large areas of Earth were covered by massive glaciers

iceberg a portion of a glacier calved off upon entering the sea

ice cap a polar cover of snow and ice

igneous rocks all rocks solidified from a molten state

impact the point on the surface upon which a celestial object has landed, creating a crater

industrialization the use of natural resources in manufacturing, transportation, and other human activities

infrared heat radiation with a wavelength between red light and radio waves

insolation all solar radiation impinging onto a planet

interglacial a warming period between glacial episodes

intertidal zone the shore area between low and high tides

intrusive any igneous body that has solidified in place below Earth's surface

island arc volcanoes landward of a subduction zone, parallel to a trench, and above the melting zone of a subducting plate

isostasy (eye-SOS-tah-see) a geologic principle that states that Earth's crust is buoyant and rises and sinks depending on its density

isotope (I-seh-tope) a particular atom of an element that has the same number of electrons and protons as the other atoms of the element but a different number of neutrons; that is, the atomic numbers are the same, but the atomic weights differ

jet stream strong winds concentrated within a narrow belt in the upper atmosphere

lahar (LAH-har) a mudflow of volcanic material on the flanks of a volcano

landfill a method of municipal solid waste disposal whereby layers of garbage are covered by layers of impermeable clay

landslide a rapid downhill movement of earth materials triggered by earthquakes and severe weather

lapilli (leh-PI-lie) small, solid pyroclastic fragments

lava molten magma that flows out onto the surface

leachate a solution created by the dissolution of soluble substances such as those found in landfills

limestone a sedimentary rock consisting mostly of calcite from shells of marine invertebrates

liquefaction (li-kwe-FAK-shen) the loss of support of sediments that liquefy during an earthquake

lithospheric a segment of the lithosphere involved in the plate interaction of other plates in tectonic activity

loess (LOW-es) a thick deposit of airborne dust

magma a molten rock material generated within Earth and that is the constituent of igneous rocks

magnitude scale a scale for rating earthquake energy

mantle the part of a planet below the crust and above the core, composed of dense rocks that might be in convective flow

mass wasting the downslope movement of rock under the direct influence of gravity

metamorphism (me-teh-MORE-fi-zem) recrystallization of previous igneous, metamorphic, or sedimentary rocks created under conditions of intense temperatures and pressures without melting

methane a hydrocarbon gas liberated by the decomposition of organic matter and a major constituent of natural gas

microearthquake a small earth tremor

midocean ridge a submarine ridge along a divergent plate boundary where a new ocean floor is created by the upwelling of mantle material.

monsoon a seasonal wind accompanying temperature changes over land and water from one season of the year to another

moraine (mah-RANE) a ridge of erosional debris deposited by the melting margin of a glacier

natural resource renewable and nonrenewable earth materials used in industrialization

nitrogen cycle the flow of nitrogen from the atmosphere to living organisms and finally back to the atmosphere when the organisms decompose

normal fault a gravity fault in which one block of crust slides down another block of crust along a steeply tilted plane

nuée ardente (NU-ee ARE-dent) a volcanic pyroclastic eruption of hot ash and gas

oil spill the dumping of crude oil from all sources onto bodies of water, which is harmful to marine life and habitats

ore body the accumulation of metal-bearing ores where the hot, hydrothermal water moving upward toward the surface mixes with cold seawater penetrating downward

ozone a molecule consisting of three atoms of oxygen in the upper atmosphere that filters out harmful ultraviolet radiation from the Sun; on the surface, it is a major component of urban smog

pahoehoe lava (pah-HOE-ay-hoe-ay) a lava that forms ropelike structures when cooled

paleontology (PAY-lee-ON-tah-logy) the study of ancient life forms, based on the fossil record of plants and animals

particulate minute particles of dust or soot disbursed in the atmosphere; human-made particulates are considered pollution

periglacial referring to geologic processes at work adjacent to a glacier

permafrost permanently frozen ground in the Arctic regions

permeability the ability to transfer fluid through cracks, pores, and interconnected spaces within a rock

petroleum a hydrocarbon fuel, including oil and natural gas, derived from ancient, buried microorganisms

photochemical a chemical reaction initiated by sunlight

photosynthesis the process by which plants produce carbohydrates from carbon dioxide, water, and sunlight

pH scale a logarithmic scale depicting the acidity or alkalinity of a substance; a pH of 0 is strongly acidic, a pH of 14 is strongly alkaline, and a pH of 7 is neutral

phytoplankton marine or freshwater microscopic, single-celled, freely drifting plant life

placer (PLAY-ser) a deposit of rocks left behind by a melting glacier; any ore deposit that is enriched by stream action

plate tectonics the theory that accounts for the major features of Earth's surface in terms of the interaction of lithospheric plates

playa (PLY-ah) a flat, dry, barren plain at the bottom of a desert basin

pollutant any substance, whether human-made or natural, that pollutes air or water

porosity the percentage of pore spaces in a rock between crystals and grains, usually filled with water
precipitation products of condensation that fall from clouds as rain, snow, hail, or drizzle; also the deposition of minerals from seawater
primary producer the lowest member of a food chain
pumice volcanic ejecta with numerous gas cavities that is extremely lightweight
pyroclastic (PIE-row-KLAS-tik) the fragmental ejecta released explosively from a volcanic vent
radioactive waste nuclear waste products from power plants, weapons manufacture, and hospital laboratories that is classified as hazardous and must be permanently disposed of under special burial conditions
reclamation a process of restoring an environment to its original condition
reef the biological community that lives at the edge of an island or continent; the shells from dead organisms form a limestone deposit
regression a fall in sea level, exposing continental shelves to erosion
reserves known and identified earth materials for immediate extraction and use
resource reserves of useful earth materials that might later be extracted
resurgent caldera a large caldera that experiences renewed volcanic activity that domes up the caldera floor
rift valley the center of an extensional spreading, where continental or oceanic plate separation occurs
rille (ril) a trench formed by a collapsed lava tunnel
riverine (RI-vah-rene) relating to a river
saltation the movement of sand grains by wind or water
salt dome an upwelling plug of salt that arches surface sediments and often serves as an oil trap
sand boil an artesian-like fountain of sediment-laden water produced by the liquefaction process during an earthquake
scarp a steep slope formed by earth movements
seafloor spreading a theory that the ocean floor is created by the separation of lithospheric plates along midocean ridges, with new oceanic crust formed from mantle material that rises from the mantle to fill the rift
seamount a submarine volcano
sedimentation the deposition of sediments
seiche (seech) a wave oscillation on the surface of a lake or landlocked sea
seismic (SIZE-mik) pertaining to earthquake energy or other violent ground vibrations
seismic sea wave an ocean wave generated by an undersea earthquake or volcano; also called tsunami
seismometer a detector of earthquake waves

shield area of exposed Precambrian nucleus of a continent

shield volcano a broad, low-lying volcanic cone built up by lava flows of low viscosity

sinkhole a large pit formed by the collapse of surface materials undercut by the solution of subterranean limestone

solifluction (SOE-leh-flek-shen) the failure of earth materials in tundra

soluble refers to a substance that dissolves in water

species groups of organisms that share similar characteristics and are able to breed among themselves

stishovite (STIS-hoe-vite) a quartz mineral produced by extremely high pressures such as those generated by a large meteorite impact

storm surge an abnormal rise of the water level along a shore as a result of wind flow in a storm

strata layered rock formations; also called beds

stratovolcano an intermediate volcano characterized by a stratified structure from alternating emissions of lava and fragments

subduction zone a region where an oceanic plate dives below a continental plate into the mantle; ocean trenches are the surface expression of a subduction zone

subsidence the compaction of sediments due to the removal of underground fluids

surge glacier a continental glacier that heads toward the sea at a high rate of advance

syncline (SIN-kline) a fold in which the beds slope inward toward a common axis

talus cone a steep-sided pile of rock fragments at the foot of a cliff

tectonics (tek-TAH-niks) the history of Earth's larger features (rock formations and plates) and the forces and movements that produce them

temperature inversion a layer of the atmosphere in which the inversion temperature increases with altitude as opposed to the normal tendency for temperature to decrease with altitude

tephra (TEH-fra) all clastic material, from dust particles to large chunks, expelled from volcanoes during eruptions

terrane (teh-RAIN) a unique crustal segment attached to a landmass

tide a bulge in the ocean produced by the Moon's gravitational forces on Earth's oceans; the rotation of Earth beneath this bulge causes the rising and lowering of the sea level generally twice daily

till sedimentary material deposited by a glacier

tillite a sedimentary deposit composed of glacial till

transform fault a fracture in Earth's crust along which lateral movement occurs; they are common features of the midocean ridges

transgression a rise in sea level that causes flooding of the shallow edges of continental margins

trench a depression on the ocean floor caused by plate subduction

tsunami (sue-NAH-me) a seismic sea wave produce by an undersea or nearshore earthquake or volcanic eruption

tuff a rock formed of pyroclastic fragments

tundra permanently frozen ground at high latitudes and elevations

typhoon a severe tropical storm in the Western Pacific similar to a hurricane

ultraviolet the invisible light with a wavelength shorter than visible light and longer than X rays

undeveloped nation a lightly industrialized, heavily populated, generally poor country

upwelling the upward convection of water currents

varves thinly laminated lake bed sediments deposited by glacial meltwater

ventifact (VEN-teh-fakt) a stone shaped by the action of windblown sand

volcanic ash fine pyroclastic material injected into the atmosphere by an erupting volcano

volcanic bomb a solidified blob of molten rock ejected from a volcano

volcano a fissure or vent in the crust through which molten rock rises to the surface to form a mountain

water pollution the contamination of water by industrial and municipal effluents

water vapor atmospheric moisture in the invisible gaseous phase

wetland land that is inundated by water and supports prolific wildlife

BIBLIOGRAPHY

THE BALANCE OF NATURE

Baskin, Yvonne. "Ecologists Put Some Life Into Models of a Changing World." *Science* 259 (March 19, 1993): 1694–1696.

Beardsley, Tim. "Tracking the Missing Carbon." *Scientific American* 264 (April 1991): 16.

Berner, Robert A., and Antonio C. Lasaga. "Modeling the Geochemical Carbon Cycle." *Scientific American* 260 (March 1989): 74–81.

Brown, Kathryn. "The Motion of the Ocean." *Science News* 158 (July 15, 2000): 42–44.

Green, D. H., S. M. Eggins, and G. Yaxley. "The Other Carbon Cycle." *Nature* 365 (September 16, 1993): 210–211.

Kerr, Richard A. "Ocean-in-a-Machine Starts Looking Like the Real Thing." *Science* 260 (April 2, 1993): 32–33.

Perkins, Sid. "An Ounce of Prevention." *Science News* 158 (July 15, 2000): 45–47.

Schell, Jonathan. "Our Fragile Earth." *Discover* 10 (October 1989): 45–50.

Schneider, Stephen H. "Climate Modeling." *Scientific American* 256 (May 1987): 72–80.

Smil, Vaclav. "Global Population and the Nitrogen Cycle." *Scientific American* 277 (July 1997): 76–81.

Zimmer, Carl. "The Case of the Missing Carbon." *Discover* 14 (December 1993): 38–39.

ENVIRONMENTAL DEGRADATION

Baskin, Yvonne. "Ecologists Put Some Life Into Models of a Changing World." *Science* 259 (March 19, 1993): 1694–1696.

Grove, Richard H. "Origin of Western Environmentalism." *Scientific American* 267 (July 1992): 42–47.

Hedin, Lars O., and Gene E. Likens. "Atmospheric Dust and Acid Rain." *Scientific American* 275 (December 1996): 88–92.

Holloway, Marguerite. "Soiled Shores." *Scientific American* 265 (October 1991): 103–116.

Homer-Dixon, Thomas F., Jeffrey H. Boutwell, and George W. Rathjens. "Environmental Change and Violent Conflict." *Scientific American* 268 (February 1993): 38–45.

King, Michael D., and David D. Herring. "Monitoring Earth's Vital Signs." *Scientific American* 282 (April 2000): 92–97.

la Riviere, J. W. Maurits. "Threats to the World's Water." *Scientific American* 261 (September 1989): 80–94.

O'Leary, Philip R., Patrick W. Walsh, and Robert K. Ham. "Managing Solid Waste." *Scientific American* 259 (December 1988): 36–42.

Pfeiffer, Beth. "Our Disposable Society." *American Legion Magazine* (January 1990): 24–25 & 58.

Raloff, Janet. "Mercurial Risks From Acid's Reign." *Science News* 139 (March 9, 1991): 152–156.

Repetto, Robert. "Accounting for Environmental Assets." *Scientific American* 266 (June 1992): 94–100.

Stolarski, Richard S. "The Antarctic Ozone Hole." *Scientific American* 258 (January 1988): 30–36.

Zimmer, Carl. "Son of Ozone Hole." *Discover* 14 (October 1993): 28–29.

CLIMATE CHANGE

Appenenzeller, Tim. "Our Fragile Climate." *Discover* 15 (January 1994): 67–69.

Broecker, Wallace S. "Chaotic Climate." *Scientific American* 273 (November 1995): 62–68.

Cohn, Jeffrey P. "Gauging the Biological Impacts of the Greenhouse Effect." *BioScience* 39 (March 1989): 142–146.

D'Agnese Joseph. "Why Has Our Weather Gone Wild?" *Discover* 21 (June 2000): 72–78.

Epstein, Paul R. "Is Global Warming Harmful to Health?" *Scientific American* 283 (August 2000): 50–57

Graedel, Thomas E., and Paul J. Crutzen. "The Changing Atmosphere." *Scientific American* 261 (September 1989): 58–68.

Horgan, John. "Arctic Meltdown." *Scientific American* 268 (March 1993): 19–28.

Karl, Thomas R., and Kevin E. Trenberth. "The Human Impact on Climate." *Scientific American* 281 (December 1999): 100–105.

Maslin, Mark. "Waiting for the Polar Meltdown." *New Scientist* 139 (September 4, 1993): 36–41.

Monastersky, Richard. "Climate's Long Lost Twin." *Science News* 157 (February 26, 2000): 138–140.

Pollack, Henry N., and David S. Chapman. "Underground Records of Changing Climate." *Scientific American* 268 (June 1993): 44–50.

White, Robert. "The Great Climate Debate." *Scientific American* 263 (July 1990): 36–43.

HYDROLOGIC ACTIVITY

Abelson, Philip H. "Climate and Water." *Science* 243 (January 27, 1989): 461.

Cathles, Lawrence M., III. "Scales and Effects of Fluid Flow in the Upper Crust." *Science* 248 (April 20, 1990): 323–328.

Garrett, Chris. "A Stirring Tale of Mixing." *Nature* 364 (August 19, 1993): 670–671).

Gleick, Peter H. "Safeguarding Our Water." *Scientific American* 288 (February 2001): 38–45.

Kerr, Richard A. "Big El Niños Ride the Back of Slower Climate Change." *Science* 283 (February 19, 1999): 1108–1109.

Macilwain, Colin. "Conservationists Fear Defeat on Revised Flood Control Policies." *Nature* 365 (October 7, 1993): 468.

Maslin, Mark. "Waiting for the Polar Meltdown." *New Scientist* 139 (September 4, 1993): 36–41.

Monastersky, Richard. "Volcanoes Under Ice: Recipe for a Flood." *Science News* 150 (November 23, 1996): 327.

———. "Seep and Ye Shall Find: Hidden Water Flow." *Science News* 149 (April 20, 1996): 245.

———. "Rivers in a Greenhouse World." *Science News* 137 (June 9, 1990): 365.

Svitil, Kathy A. "Groundwater Secrets." *Discover* 16 (March 1996): 28.

Williams, Nigel. "Dams Drain the Life Out of Riverbanks." *Science* 276 (May 2, 1997): 683.

COASTAL PROCESSES

Flanagan, Ruth. "Sea Change." *Earth* 7 (February 1998): 42–47.

Folger, Tim. "Waves of Destruction." *Discover* 15 (May 1994): 68–73.

Fritz, Sandy. "The Living Reef." *Popular Science* 246 (May 1995): 48–51.

Gonzalez, Frank I. "Tsunami!" *Scientific American* 280 (May 1999): 56–65.

Kerr, Richard A. "From One Coral Many Findings Blossom." *Science* 248 (June 15, 1990): 1314.

Lockridge, Patricia A. "Volcanoes and Tsunamis." *Earth Science* 42 (Spring 1989): 24–25.

Monastersky, Richard. "Against the Tide." *Science News* 156 (July 24, 1999): 63.

Norris, Robert M. "Sea Cliff Erosion: A Major Dilemma." *Geotimes* 35 (November 1990): 16–17.

Parfit, Michael. "Polar Meltdown." *Discover* 10 (September 1989): 39–47.

Pendick, Daniel. "Waves of Destruction." *Earth* 6 (February 1997): 27–29.

Svitil, Kathy A. "The Sound and the Fury." *Discover* 16 (January 1995): 75.

TECTONIC HAZARDS

Bolt, Bruce A. "Balance of Risks and Benefits in Preparation for Earthquakes." *Science* 251 (January 11, 1991): 169–174.

Horgan, John. "Volcanic Disruption." *Scientific American* 26 (March 1992): 28–29.

Kerr, Richard A. "Seismologists Learn the Language of Quakes." *Science* 271 (February 16, 1996): 910–911.

———. "Volcanoes: Old, New, and—Perhaps—Yet to Be." *Science* 250 (December 21, 1990): 1660–1661.

Milstein, Michael. "Cooking up a Volcano." *Earth* 7 (April 1998): 24–30.

Mouginis-Mark, Peter J. "Volcanic Hazards Revealed by Radar Interferometry." *Geotimes* 39 (July 1994): 11–13.

Normile, Dennis. "A Wake-up Call from Kobe." *Popular Science* (February 1996): 64–68.

Pendick Daniel. "Himalayan High Tension." *Earth* 5 (October 1996): 46–53.

Simkin, Tom. "Distant Effects of Volcanism—How Big and How Often." *Science* 264 (May 13, 1994): 913–914.

Swinbanks, David. "Volcano Prediction Problems." *Nature* 351 (June 13, 1991): 511.

Tilling, Robert I., and Peter W. Lipman. "Lessons in Reducing Volcano Risk." *Nature* 364 (July 22, 1993): 277–280.

Wakefield, Julie, and Daniel Pendick. "Earthquake Reality Check." *Earth* 4 (August 1995): 20–21 & 60–61.

LOSING GROUND

Davidson, Keay. "Yosemite Landslide Caught in the Act." *Earth* 5 (December 1996): 12–13.

Friedman, Gerald M. "Slides and Slumps." *Earth Science* 41 (Fall 1988): 21–23.
Glanz, James. "Erosion Study Finds High Price for Forgotten Menace." *Science* 267 (February 24, 1995): 1088.
Kerr, Richard A. "Volcanoes With Bad Hearts Are Tumbling Down All Over." *Science* 264 (April 29, 1994): 660.
Monastersky, Richard. "Spotting Erosion from Space." *Science News* 136 (July 22, 1989): 61.
———. "Soil May Signal Imminent Landslide." *Science News* 134 (November 12, 1988): 318.
Nuhfer, Edward B. "What's a Geologic Hazard?" *Geotimes* 39 (July 1994): 4.
Pendick, Daniel. "Ashes, Ashes, All Fall Down." *Earth* 6 (February 1997): 32–33.
Peterson, Ivars. "Digging Into Sand." *Science News* 136 (July 15, 1989): 40–42.
Pinter, Nicholas, and Mark T. Brandon. "How Erosion Builds Mountains." *Scientific American* 276 (April 1997): 74–79.
Reganold, John P., Robert I. Papendick, and James F. Parr. "Sustainable Agriculture." *Scientific American* 262 (June 1990): 112–120.
Shaefer, Stephen J., and Stanley N. Williams. "Landslide Hazards." *Geotimes* 36 (May 1991): 20–22.
Zimmer, Carl. "Landslide Victory." *Discover* 12 (February 1991): 66–69.

DESERTIFICATION

Cambell, John M. "Desert Extremes." *Earth* 7 (April 1998): 39–46.
Hedin, Lars O., and Gene E. Likens. "Atmospheric Dust and Acid Rain." *Scientific American* 275 (December 1996): 88–92.
Mack, Walter N., and Elizabeth A. Leistikow. "Sands of the World." *Scientific American* 275 (August 1996): 62–67.
Madeley, John. "Will Rice Turn the Sahel to Salt." *New Scientist* 140 (October 9, 1993): 35–37.
Martin, John. "Desert Extremes." *Earth* 8 (April 1998): 39–46.
Monastersky, Richard. "Sahara Dust Blows Over United States." *Science News* 148 (December 23 & 30, 1995): 431.
Nori, Franco, et al. "Booming Sand." *Scientific American* 277 (September 1997): 84–89.
Pennisi, Elizabeth. "Dancing Dust." *Science News* 142 (October 3, 1992): 218–220.
Raloff, Janet. "Holding on to the Earth." *Science News* 126 (October 30, 1993): 280–281.
Reganold, John P., Robert I. Papendick, and James F. Parr. "Sustainable Agriculture." *Scientific American* 262 (June 1990): 112–120.

Sen, Amartya. "The Economics of Life and Death." *Scientific American* 268 (May 1993): 40–47.

Szelc, Gary. "Where Ancient Seas Meet Ancient Sand." *Earth* 5 (December 1997): 78–81.

Zimmer, Carl. "How to Make a Desert." *Discover* 16 (February 1995): 51–56.

NATURAL RESOURCES

Abelson, Philip H. "Increased Use of Renewable Energy." *Science* 253 (September 1991): 1073.

Barnes, H. L., and A. W. Rose. "Origins of Hydrothermal Ores." *Science* 279 (March 27, 1998): 2064–2065.

Brimhall, George. "The Genesis of Ores." *Scientific American* 264 (May 1991): 84–91.

Corcoran, Elizabeth. "Cleaning Up Coal." *Scientific American* 264 (May 1991): 107–116.

Davis, Ged R. "Energy for Planet Earth." *Scientific American* 263 (September 1990): 55–62.

Fulkerson, William, Roddie R. Judkins, and Manoj K. Sanghvi. "Energy from Fossil Fuels." *Scientific American* 263 (September 1990): 129–135.

Hapgood, Fred. "The Quest for Oil." *National Geographic* 176 (August 1989): 226–263.

Hubbard, Harold M. "The Real Cost of Energy." *Scientific American* 264 (April 1991): 36–42.

Kerr, Richard A. "Geothermal Tragedy of the Commons." *Science* 253 (July 12, 1991): 134–135.

Rosenberg, A. A., et al. "Achieving Sustainable Use of Renewable Resources." *Science* 262 (November 5, 1993): 828–829.

Rosenfield, Arthur H., and David Hafemeister. "Energy-Efficient Buildings." *Scientific American* 258 (April 1988): 78–85.

Suess, Erwin, et al. "Flammable Ice." *Scientific American* 281 (November 1999): 76–83.

LAND USE

Brown, Barbara E., and John C. Ogden. "Coral Bleaching." *Scientific American* 268 (January 1993): 64–70.

Clark, William C. "Managing Planet Earth." *Scientific American* 261 (September 1989): 47–54.

Diamond, Jared. "Playing Dice with Megadeath." *Discover* 11 (April 1990): 55–59.

Ehrlich, Paul R. "Facing the Habitability Crises." *BioScience* 39 (July/August 1989): 480–482.

Goulding, Michael. "Flooded Forests of the Amazon." *Scientific American* 266 (March 1993): 114–120.

Holloway, Marguerite. "Population Pressure." *Scientific American* 267 (September 1992): 32–38.

Horgan, John. "Up in Flames." *Scientific American* 264 (May 1991): 17–24.

May, Robert M. "How Many Species Inhabit the Earth?" *Scientific American* 267 (October 1992): 42–48.

Repetto, Robert. "Deforestation in the Tropics." *Scientific American* 262 (April 1990): 36–42.

Ruckelshaus, William D. "Toward a Sustainable World." *Scientific American* 261 (September 1989): 166–174.

Schell, Jonathan. "Our Fragile Earth." *Discover* 10 (October 1989): 45–50.

Schmidt, Karen. "Life on the Brink." *Earth* 6 (April 1997): 26–33.

Terborgh, John. "Why American Songbirds Are Vanishing." *Scientific American* 266 (May 1992): 98–104.

INDEX

Boldface page numbers indicate extensive treatment of a topic. *Italic* page numbers indicate illustrations or captions. Page numbers followed by *m* indicate maps; *t* indicate tables; *g* indicate glossary.

D